ROUTLEDGE LIBRARY EDITIONS:
SOCIAL AND CULT

Volume 5

THE SOCIAL GEOGRAPHY OF
MEDICINE AND HEALTH

THE SOCIAL GEOGRAPHY OF MEDICINE AND HEALTH

JOHN EYLES
AND KEVIN J. WOODS

Routledge
Taylor & Francis Group

LONDON AND NEW YORK

First published in 1983

This edition first published in 2014
by Routledge
2 Park Square, Milton Park, Abingdon, Oxfordshire OX14 4RN

and by Routledge
711 Third Avenue, New York, NY 10017

First issued in paperback 2016

Routledge is an imprint of the Taylor & Francis Group, an informa business

British Library Cataloguing in Publication Data
A catalogue record for this book is available from the British Library

ISBN: 978-0-415-83447-6 (Set)
ISBN-13: 978-1-138-99810-0 (pbk)
ISBN-13: 978-0-415-73321-2 (hbk)

Publisher's Note
The publisher has gone to great lengths to ensure the quality of this reprint but points out that some imperfections in the original copies may be apparent.

Disclaimer
The publisher has made every effort to trace copyright holders and would welcome correspondence from those they have been unable to trace.

The Social Geography of
of
Medicine and Health

John Eyles and Kevin J. Woods

CROOM HELM
London & Canberra

ST. MARTIN'S PRESS
New York

© 1983 J.D. Eyles and K.J. Woods
Croom Helm Ltd, Provident House, Burrell Row,
Beckenham, Kent BR3 1AT

British Library Cataloguing in Publication Data

Eyles, John
  The social geography of medicine and health.
  1. Social medicine
  I. Title II. Woods, Kevin J.
  306'.4   RA418
ISBN 0-7099-0257-3

All rights reserved. For information, write:
St. Martin's Press, Inc., 175 Fifth Avenue, New York, N.Y. 10010
First published in the United States of America in 1983

Library of Congress Cataloging in Publication Data

Eyles, John.
  The social geography of medicine and health.

  Bibliography: p.
  Includes index.
  1. Medical geography. 2. Social medicine.
I. Woods, Kevin, J. II. Title.
RA792.E94   1983   362.1'042   83-2921
ISBN 0-312-73292-9

Typeset by Leaper & Gard Ltd, Bristol
Printed and bound in Great Britain by
Biddles Ltd, Guildford and King's Lynn

CONTENTS

FIGURES

TABLES

PREFACE

This book is primarily an exposition of medicine and health from a social geographical perspective. We make no apology for starting from this framework or for going beyond it in trying to explain and understand the phenomena under investigation. We feel that spatial patterns formed by social phenomena are a good starting point for examining the relationships between medicine, health and society. Many health conditions vary significantly across space. Most health care provisions, at least of the public variety, are allocated between and within territorial units. Further, social geography, along with its parent discipline, has a powerful synthesising nature which assists the search for meaning and explanation. We address, however, the problem of different perspectives on health and health care in Chapter 1. But to anticipate, we see social geography, partly because of its subject-matter and partly because of the academic climate in which it finds itself, as being part of the broader social science concern with the dissolution of academic boundaries in attempting to understand the social world. We feel unconstrained in selecting elements from different perspectives to assist us with these endeavours.

Medicine and health are undeniably social phenomena. Good health is a necessary precondition for successful functioning in any type of society. Such a statement should be taken to include social as well as biological functioning. Good health is needed for work, leisure and social activity. It enhances general quality of life, whereas poor health or disease both create and are created by deleterious social and economic conditions. Medicine, as one basis of systems of care and as one human response to disease, is also socially produced. Medicine and health care are not simply socially neutral technical procedures for combating disease and ill-health. They do combat such problems and we shall examine the geographical contribution to understanding the relationship between man and disease, mainly as mediated by environment. But they are also social products in the sense that the containing society significantly moulds and influences them. Indeed, we shall argue that it is not only medical care that is so shaped, but our very conceptions of health and illness. We thus see medicine and health as being truly embedded in the social system with the shape of that system significantly affecting the definition of health and the nature of health care provision. We do not attempt to suggest that there is some inevitable sequence of developments that simply follows some logic of industrialism. Culture, history and economy ensure that different societies have different definitions and different care systems.

We shall, however, investigate the geographical work which attempts to use and apply mathematical and theoretical models to the development of health care. In a way, such attempts are suggesting that similar ideas and practices can be applied in different social settings. These economically-oriented perspectives are challenged by considerations of behavioural factors and of existing, operating economic and political systems. Indeed, we consider this embedding of medicine and health in the societal order to be of central importance. We shall emphasise this point on several occasions. There is also some overlap of material in several of the chapters. This overlap – or linking – is in fact a key element in our presentation. We hope to provide a sequential argument which moves, after a discussion of perspectives and definitions (i.e. the background), from medicine as seen through an understanding of disease environments and health problems (i.e. the raw material), to health care policy theoretically and practically conceived (i.e. the response) to society as the significant mediation between man and his conception and treatment or ill-health (i.e. the context). Thus environment and culture are necessary to understand medicine and health but such understanding must also be predicated on the dialectical relationship between man and society. It is our hope that we address some of the elements of this relationship – a relationship in which ideas (conceptions) are as important as actions (policies) and which sees social geography as a necessary part in explanation.

ACKNOWLEDGEMENTS

All academic endeavours build on that which is already present. This book is no exception, but in working in the field of medicine and health care we have been able to draw on a vast and rich existing literature in geography, sociology, epidemiology and political studies. We have been fortunate in other ways too. At Queen Mary College, we are members of a thriving Health Research Group, consisting of staff members and graduate students in the Department of Geography. We wish to thank our colleagues in the Group for providing a thought-provoking environment in which to work. In particular we are indebted to David Smith, not only for providing a stimulating academic atmosphere in the most pressing of times, but also for giving us the incentive and encouragement to begin and finish the book. We are also grateful for the many valuable comments he made on reading an earlier version of the manuscript. Similarly, Eva Alberman of the Department of Clinical Epidemiology in The London Hospital Medical College has been an invaluable source of quiet advice and encouragement. Parts of the book were written whilst both of us held posts as visiting lecturers in Australia and we wish to record our thanks to the many friends we made at the University of New England, Armidale and Flinders University, Adelaide who gave us the opportunity to gather our thoughts and commit some of them to paper. These opportunities enriched our own collaboration. While we each had individual responsibilities, with John Eyles Chapters 1, 2, 3 and 7, Kevin Woods 4 and 5 and both different parts of Chapter 6, all parts were read by both of us and in several places substantially modified.

We would also like to acknowledge the diligence and perseverance of secretarial and technical staff in the Department of Geography, QMC, the Department of Clinical Epidemiology, LHMC and outside. Original figures were drawn by the cartographic unit at QMC under the direction of Lynne Fraser, while Pat Mitchell made sense of our lists of references. Working against the pressures of time and virtually unreadable writing, Linda Agombar, Eileen Bruce, Dorothy Eyles, Linda Goodchild, Carol Gray and Sue Nettleton produced a high quality manuscript. Many thanks to one and all. Finally, we dedicate this book to our long suffering but understanding families.

1 PERSPECTIVES ON HEALTH AND HEALTH CARE

It is not the purpose of this chapter to provide a review of all the many perspectives from which health and health care have been approached. We wish to explore briefly those perspectives that seem relevant and important to our own purposes, namely explicating the limitations and assumptions of present geographical endeavour and attempting to understand health and health care in terms of the definitions, meanings and social contexts by and in which such phenomena appear.

The Biomedical Perspective

It is perhaps true to say that the biomedical model is so deeply interwoven with ways of thinking about and working in medicine that it is often forgotten that this is but one perspective. It is often regarded, therefore, as the sole representation of reality. But like other perspectives, it defines, classifies and assesses relationships among phenomena in a particular way. Thus, it presupposes that there are specific disease entities each associated with specific biological processes. Aetiology is thus seen as biologically specific.

Mishler (1981) identifies four major assumptions of the biomedical perspective. First, disease is defined as deviation from normal biological functioning. Thus Cohen (1961, 169) suggests that 'disease indicates deviations from the normal – these are its symptoms and signs'. Engel (1963) suggests that health involves an individual maintaining a balance so as to be reasonably free of pain and disability. Notions of balance and normality raise questions about the standards against which such assessments are made. Such questions are addressed more fully in Chapter 2. Suffice it to say here that workers within the biomedical perspective have begun to question the accepted meaning of normality. Redlich (1957), for example, asks normal for what and for whom, while Ryle (1961) suggests that even in physiological phenomena, variation is so constantly at work that no rigid pattern is discernible.

Secondly, there exists the doctrine of specific aetiology (see Dubos, 1960). The doctrine – the basis of modern Western medicine – suggests a movement from initial classification based on specific symptoms through a clustering of such symptoms into syndromes to identifying the specific pathogenesis and pathology of diseases. In particular, the presence of micro-organisms or disease vectors (see Chapter 3) seemed to explain the occurrence of disease. The bacterial origin of infectious disease, discovered by Pasteur, and the role of agents in epidemics, identified by Koch, pointed to the importance of medical interventions to eradicate such problems. Indeed, the role of scientific medicine cannot be overemphasised, not only in identifying the sources of disease and in providing preventive and therapeutic measures, but in also contributing to advances in hygiene (see McKeown, 1979). But, as Dubos (1963) observes, disease is rare given the number of parasites present in the environment. Disease only usually results when natural resistances have been undermined by the stresses and strains of life. Such stresses may in their turn be dependent on available coping mechanisms (Mechanic, 1978), especially the presence of informal, localised social support networks (Cobb, 1976; see Chapter 7). Thus, the doctrine of specific aetiology provides a simple cause and effect model which cannot adequately explain why specific individuals in specific places are sick or ill (see Chapter 3).

Thirdly, there is the assumption of generic diseases. In other words, 'each disease has specific and distinguishing features that are universal to the human species. That is, disease symptoms and processes are expected to be the same in different historical periods and in different cultures and societies' (Mishler, 1981, 9). It is, of course, important to recognise that biological events are observable, measurable and orderly. But there is an important distinction between a description of a biological process and definition of symptoms of illness. Western scientific medicine is only one way of defining illness. Lay concepts also define and, as we shall see in Chapter 6, there is much cultural variation in definitions of sickness and care. As Fabrega (1975, 969) argues Western medicine 'thus constitutes our own culturally specific perspective about what disease is, and how medical treatment should be pursued; and like other medical systems, biomedicine is an interpretation which "makes sense" in the light of cultural traditions and assumptions about reality'.

Fourthly, there is the assumption of the scientific neutrality of medicine. Doctors see themselves as bioscientists and adopt the objectivity and neutrality of the scientific method. Medicine is thus seen as an isolated and independent section of society. Such a view has many ramifications for how medicine is seen and for how the relationships within medical practice are managed. Because of the importance of this assumption and its implied suggestion that medicine is divorced from social, economic and political concerns, we shall examine this topic in detail in Chapter 2. Suffice it to say, that the dislocation of medicine stems from the peculiar place of science in Western medicine. In small-scale societies this dislocation is not present. As Fabrega (1976, 290) suggests:

in a logical sense, disease among nonliterates is directly tied to the social behaviour of the person and to his ability to function. All types of disease raise social and personal questions about the individual and his immediate group. Thus, disease and medical care are directly woven into the social fabric. In our culture science has provided us with disease forms which, in logical grounds, are not connected to the social fabric.

This mediation between ill-health and environment by technological, professional and cultural practices will be a key element structuring Chapter 6.

It is possible to identify a fifth assumption of the biomedical perspective in that medical knowledge advances in a linear fashion, solving one problem for the entire population of sufferers before moving on to tackle yet another. This idea of sequential progress makes light of different predispositions to disease, different attitudes to sickness and differential access to medical resources dependent on differential power and advantage. It also assumes that financial resources are available to allow the technical solutions to be developed and applied to medical problems. In other words, it fails to identify competing wants and limited resources. It is essentially non-economic.

The Economic Perspective

Culyer (1976, viii) suggests that 'economics, as the science of choice, can lay bare the necessary elements in making social choices about meeting needs for health care, and its techniques can also provide empirical estimates of the dimensions of some of the relevant concepts . . . Economics is the integrative element.' The economic perspective recognises the scarcity of resources and the multiplicity of wants and values; to determine the fair and efficient allocation of health resources it is necessary to examine these wants and values.

The notion of need is problematic in the economic perspective. Culyer *et al.* (1972) suggest that 'need' should be removed from public policy as it is ambiguous with value implications and unclear cost implications. Williams (1974) further argues that as need is an evaluative and normative concept, it should be related to the end or goal being sought. In that way, need for health services is seen as a technical and economic matter: a matter for the providers of health services. It is not necessarily related to the individual's own predisposition or choice. This perspective argues, therefore, that:

> the need for health care is defined by reference to some third party's view as to what a particular individual or class of individuals themselves *ought* to receive. The demand for health care, however, is indicated by the individuals themselves in making claims upon health care resources. (Culyer, 1976, 19)

While the economic perspective may make passing reference to some of the problems of lower income groups (dealt with in Chapter 3), it is dominated by the production of demand rather than the distribution of need and care. Smith (1977) points to the reluctance of economists to deal explicitly with questions of distribution until quite recently because of the ethical nature of interpersonal-utility comparisons. Indeed, the conventional treatment of distribution, seen in Culyer's work, is based on Pareto optimality which exists when it is impossible to make anyone better off without at the same time making someone else worse off. Nothing is said about the existing distribution of health care. If it is highly unequal to start with, a welfare 'improvement' in Pareto's terms may see the rich and healthy improve their lot while the poor and sick remain as they are. It is possible to argue that the private provision of health care in societies previously dominated by a state-provided system may represent such a condition (see Chapter 6).

Demand for health is seen as the crucial factor in this perspective. Such a demand is regarded as a demand for an investment good that yields services over a period of time. The demand is determined in part by the individual's particular stock of health – a capital good which depreciates over time, but which can be increased by investment. The relationship between investment and the stock of health is dependent on the amount of time and money devoted to improving the stock. The key variables in investment – diet, exercise, housing, consumption habits, environmental factors, education and information – are under the variable control of the individual. If they are under his control then the individual chooses his preferred stock of health. As age increases and 'costs' rise, a lower stock is chosen. If income rises, investment is likely to rise, although some expenditures – alcohol, cars – may be harmful. Individual preference and consumer sovereignty are seen to reign – a triumph for neo-classical economic theory – and the market, the supplier of health care, responds to these consumer pressures.

From the economic perspective, health becomes a matter of the deliberate consumption of goods and services. Such consumption is seen as an investment in the stock of health of the individual. It is also regarded as being shaped entirely by individual preferences. 'The basic value judgement . . . is that the supply of goods and services, including medical care, should as nearly as possible be based upon individual preferences' (Lees, 1961, 14). As Donnison (1975, 423) argues:

> To rely wholly on evidence about short-run individual preferences as the criterion for collective action is *unhistorical*, because it takes too little account of the way in which preferences are shaped (and could be reshaped) by influences extending over long periods of time. It is *unsociological* because it treats people as atomistic individuals, deciding only for themselves, rather than as members of classes, families and other groups which support and constrain them (and could influence them in different ways if the social structure changed).

Such a conception, which we exemplify in critical detail in Chapters 4 and 5, also treats demand simply as a manifest phenomenon. It conveniently shelves the question of need which may or may not be manifested as demand. Demand is simply an individualistic response (see also Chapter 2), divorced from its social and economic context. Both demand and need must be seen as being societally derived. Indeed it is necessary to relate need, want and demand to each other. Smith (1977) suggests that a need implies an imperative while want suggests the source of an acquisitive desire. Bradshaw (1972) offers us four distinct definitions of need which help to relate the three terms: first, normative need based on bureaucratic determination and the remedial provision of minimum levels of adequacy – the basis of many means-tested health programmes (see Chapter 2); secondly, felt need, the stated wants of those for whom services are offered – the closest to the economic perspective and the basis of the location – allocation models discussed in Chapter 4; thirdly, expressed need – the lack of a good or service may provoke action or demand which may or may not be effective; and fourthly, comparative need, in which comparisons with others leads to a general recognition that a service is necessary.

Needs, wants and demands are, therefore, essentially social. Such phenomena in the field of health and health care are not simply a question of consumption satisfiable by market principles. They also depend on many intangible goods and bads that shape people's lives. They depend, too, on other individuals and groups and how their perceptions and experiences interact one with another. Further, the needs and demands for health care are related to the type of society that contains them. In capitalist society, the satisfaction of needs often means the possession of objects. These needs often become impoverished and homogenised. Acquisitive desire (want), manifested as demand, appears to dominate and the satisfaction of such wants may help commoditise social relations and the social fabric (see Chapter 2). Further, the atomisation of the individual and his designation as a pre-social entity cannot be overcome within the economic perspective. Such considerations require an assessment of the sociality of man.

The Behavioural Perspective

It is possible in the social sciences in general to see the behavioural perspective as a corrective to the economic in that the former treats the individual as a social entity explicitly. During the late 1950s and early 1960s there was a surge of interest in behavioural studies. Economics began to lose its primacy as *the* explanatory social science. The notions of economic man and consumer sovereignity and the goal of maximising satisfaction (or welfare) began to be questioned. Increasing importance came to be attached to social and psychological variables in analysing human activity (see Eyles and Smith, 1978). The conventional conception of economic man fails then to account for much of the variation in individual behaviour. Such variation partially explains why the market fails to operate along the lines of neo-classical economic theory. Indeed, as Simon (1957, 198) argues;

> The capacity of the human mind for formulating and solving complex problems is very small compared with the size of the problems whose solution is required for objectively rational behaviour in the real world – or even for a reasonable approximation to such objective rationality.

Simon suggests, therefore, that individuals 'satisfice' rather than 'optimise'. In other words, they seek to discover and select satisfactory alternatives. An outcome – health status, access to health facilities – is judged to be satisfactory or unsatisfactory according to the aspiration level and stock of knowledge of the individual.

The behavioural perspective thus focuses attention on the individual, his social attributes and aspirations. In health and health care research, consideration of the individual is mainly seen in doctor–patient interaction studies. Most such studies have addressed the problem of communication in such interactions (see Korsch and Negrete, 1972; Waitzkin and Stoeckle, 1976). It has been shown that problems in communication – information disclosure, adequate explanation, etc. – vary according to the age, race and occupational status of the patient. Variations in communication also occur according to the medical facility attended – emergency room, outpatient clinic or inpatient ward. Other important studies in this area concern patient compliance: whether the patient follows the doctor's recommendations. Dunbar and Stunkard (1979) discovered that between 20 and 82 per cent of patients do not follow the recommended regimens. It appears that patients are less compliant when there has been limited information exchange, dissatisfaction with the interview and restricted responsiveness by the doctor. In other words, the interaction depends not only on medical setting and patient attributes but also on the expectations taken to the setting by the patient. We introduce such studies here simply because they broaden the behavioural perspective to include consideration of the meanings and definitions taken to the health setting by the individual. Such meanings and the contexts that produce them are considered below.

The behavioural perspective has also been influential in medical geography and the assessment of the location of health facilities. Haynes and Bentham (1979) examined the influence of such individual attributes as age, sex, occupation status and personal mobility on the utilisation of community hospitals in rural England. In their study of Swansea, Herbert and Peace (1980) demonstrated how personal mobility affects the satisfaction of the elderly with community facilities. In a study of primary care – doctors' surgeries in West Glamorgan – Phillips (1981) showed how social status, personal mobility and previous residence influence the surgery attended. In the case of the latter variable, many patients wish to remain with their doctor even when they move. Phillips (1979) termed such movement relict patterns of travel. He found that age of respondent and the presence of pre-school children in the household had little influence on surgery attendance. Such studies demonstrate the importance of the behavioural perspective in that they challenge the predictions of neo-classical economic theory and also central place theory (see also Chapter 4). Conceptualisation of the individual as a bundle of social attributes may present difficulties. It is not easy to obtain a sense of the social context from which the individual derives. Without that sense, it is possible to see one form of behaviour as 'normal' – that predicted by theory, or that practised by the vast majority of individuals – and see other behaviours as maladaptive, as deviations from the normal. Such a view has parallels in the biomedical perspective. The problems of concentrating on the individual are examined in Chapter 2. As noted in the previous paragraph, an attempt to solve this problem has involved consideration of social meanings and context of the individual.

Such consideration is encompassed by broadening the behavioural framework to include the interpretive and phenomenological traditions in which individual attributes are no longer taken as given. They are seen as being socially produced. Thus Berger and Luckmann (1967, 49) argue that 'While it is possible to say that man has a nature, it is more significant to say that man constructs his own nature, or more simply, that man produces himself.' The relationship between man and his social world is a dialectic one, that is they interact with each other. In the health field, such a perspective has been primarily concerned with the social construction of diagnoses and illnesses, that is the meanings of diagnosis and illnesses. Studies have concentrated on the societal reaction to specific disorders and diseases rather than on how individuals come to obtain their definitions. In many ways, the production of such definitions by family, social network, local neighbourhood and class are of greater interest to the social geographer than societal reaction, emphasising as they do the relationships between local and national meaning-systems and their impacts on attitudes and behaviour (see Chapter 7). The notion of societal reaction is based on labelling theory which suggests that definitions of and responses to deviance (illness) are a function of social rules (see Schur, 1971). Waxler (1981) demonstrates how these 'rules' operate in the case of leprosy. She argues that the definition of the disease and associated social expectations depend as much on society and culture ('the rule-producers') as on the biological characteristics of the disease itself. These rules define in fact a process of negotiation. Individuals diagnosed as suffering from leprosy have to learn how to be such sufferers by negotiating with friends and relations as well as with those in the health care system. The importance of such negotiation and definitions of disease can be seen by looking at the variations in the treatment of sufferers. In India, they are isolated and treated as outcasts; in Sri Lanka, they often remain with their families but stigmatise themselves by not going out or visiting; in Nigeria, they remain as active members of the community unless the disease is well advanced: even then their fate-begging is an accepted and non-stigmatised role.

It is important to recognize that the rules surrounding (and moral definitions of) all illnesses and diseases vary from society to society. Such definitions affect the behaviour of sufferers and the practice of health professionals. Thus the 'attributes' of individuals and the values of doctors do not lie outside society. Individuals and medical institutions are part of society itself (see Chapter 2). The rules and definitions arise, then, from particular medical/social/historical circumstances and become embedded in the attitudinal and institutional structures of society. These circumstances and structures are worthy of analysis in their own rights; such analysis is in fact an essential feature of any attempt fully to understand health.

The Materialist Perspective

The materialist perspective is broadly conceived here to cover those studies that see the social, economic and political structures of society as major factors in shaping health care and health care provision. It is possible to identify three strands in the perspective. First, there is that element that examines the institutional structure of health care, and relates that structure to industrialism. Illich (1974) argues that industrialism is the main force shaping our societies and that industrial expansion damages irreparably sectors like health and education. Indeed, such expansion produces frustrations. These 'have become manifest from private-enterprise systems and from socialised care [and] have come to resemble each other frighteningly' (Illich, 1974, 921). Illich thus suggests that there is a 'logic' of industrial development that similarly affects all types of society (see Chapters 2 and 7).

The industrialisation of medicine entails its professionalisation and bureaucratisation. The major conflicts thus ensue between medical bureaucracies and the consumers of the products of these bureaucracies. Such conflict appears as iatrogenesis, that is damage done by the provider. There are three major forms of iatrogenesis:

> *clinical*, when pain, sickness and death result from the provision of medical care; . . . *social*, when health policies reinforce an industrial organisation which generates dependency and ill-health; and . . . *structural*, when medically sponsored behaviour and delusions restrict the vital autonomy of people by undermining their competence in growing up, caring for each other and ageing. (Illich, 1975, 165)

To overcome these problems, Illich suggests greater individual responsibility for care and the deprofessionalisation and debureaucratisation of medicine.

A powerful critique of Illich is provided by Navarro (1976) who suggests that social iatrogenesis is not caused by the manipulation of people by medical bureaucracies. Such manipulation is itself a symptom of the basic needs of the economic and social institutions of capitalist societies. The medical bureaucracies reinforce and capitalise on what is already there - the need for consumption, a dependency of the individual on something that can be bought Such dependency – 'commodity fetishism' (see Chapter 2) – i intrinsically necessary for the survival of a system based on commodity production. Medical bureaucracies are not the generators but the administrators of these necessary dependencies and consumptions. Further, with structural iatrogenesis, Navarro argues that the medical bureaucracies are again servants of a higher category of power – the dominant class. Indeed, one of the functions of the service bureaucracies (including medicine) is to legitimise and protect this class and its power relations. An important element in such protection is the channelling of dissatisfaction, social control; that is that which Illich calls structural iatrogenesis.

Illich's view is, therefore, seen as profoundly conservative. We assess in the next chapter the rising tide of individual responsibility in health care. Illich ignores the power and class relations of society. Thus in attempting to tackle the institutional, he fails to consider the nature of the social formation in affecting conceptions of health and the delivery of health care. We thus turn to the second strand of this perspective, which sees the structural determinants of economy and polity as vital in shaping the nature of health and health care. Thus Doyal (1979b) argues that many illnesses arise from the conditions of work under capitalism, while Navarro (1976) attempts to show that the same economic and political forces that determine the class structure of the United States also determine the nature and functions of the US health sector:

> Indeed, the composition, nature and functions of the latter are the result of the degree of ownership, control and influence that primarily the corporate and the upper middle classes have on the mean of production, reproduction and legitimisation of US society. (Navarro, 1976, 164)

Such a view is supported by Elling (1981) whose view of health care provision is predicated on Mandel's (1975) functions of the capitalist state. Mandel argues that there are three major functions, namely: the provision of the general conditions of production which cannot be assured by the private activities of the dominant class; the repression of threats to the mode of production by the dominated class by the army, policy and prison; and the integration of the dominated class to ensure the pre-eminence of ruling class ideology and their acceptance of subservience. Health care is thus seen as an element in social reproduction and of social integration. Health expenditure is seen as reducing dissent and reproducing capitalist relations in the capitalist core and as helping create a stable investment environment in peripheral nations (see Brown 1976).

In this strand, therefore, capitalist relations are seen as determinate entities, shaping the nature of society and of health care. Such a view is important. It can demonstrate the structural implications of a surface phenomenon like health care. Thus Aronson (1978) in her critique of a study of schistosomasis in St Lucia points to the neglected dimensions of economic change and development and their impact on the rates of the disease. Aronson, in fact, examines the relationship between dominant economic interests and the medical conception of illness and its socially relevant consequences. Although schistosomasis is an occupational disease of banana plantation workers, it does not reduce productivity or affect the economic interests of plantation owners. Affected workers adjust to the disease by working longer hours, which reduces leisure-time and drains energy. Such coping is functional for the existing distribution of power and, indeed, the workers' lack of political power means that the plantation owners need not make the changes – a shift to smaller-scale production – that would reduce both the incidence of the disease and their own economic power.

Such insights are important, but there is an overall concern that the structural view of the world not only is economically deterministic but also suggests that capitalist relations are inevitably reproduced. While this is not the place to examine in detail the problems of structuralism (see Eyles, 1981a), the approach does inform parts of Chapter 2. The latter part of that chapter and Chapter 7 address the problem of inevitability. In the meantime, we wish to tackle the problem of economic determinism and turn to the third strand of the materialist perspective – the cultural critique of medicine and society. Such an approach looks not only at the economic and technical relationships found in medicine but also at the social-cultural relationships which express and reinforce the social relations of the containing society. Thus, although carried out from a materialist perspective, ideological or hegemonic forces are central to such work.

By hegemony, Gramsci meant the permeation throughout civil society – including a whole range of structures and activities like trade unions, schools, the churches and the family – of an entire system of values, attitudes, beliefs, morality etc. that is in one way or another supportive of the established order and the class interests that dominate it. Hegemony in this sense might be defined as an 'organising principle', or world-view, that is diffused by agencies of ideological control and socialisation into every area of daily life. (Boggs, 1976, 39)

Medicine is one such agency, the ends of which are the effective self identification of the individual with the hegemonic forms – 'a specific and international "socialisation" which is expected to be positive but which, if that is not possible, will rest as a resigned recognition of the inevitable and the necessary' (Williams, 1977, 118).

The impact and effect of such hegemony and socialisation is examined by Ehrenreich and English (1978). In the nineteenth century, the idea of sickness pervaded upper- and upper-middle-class female culture. Doctors saw women as being innately sick, and although this did not of course make them sick, it provided a rationale against allowing women to act in any other way. The recommended cure for 'normal illnesses' – menstruation, pregnancy and menopause – was bed rest, an effective way of excluding women from economic and political life. Thus, medical ideas, in hegemonic form, work with dominant interests, upper-class males, to ensure the continuation of those interests not in a coercive way but in an apparently caring fashion. In this example, we can see the distinction between the behavioural and materialist perspectives. It is not overtly concerned with the motivation of individual doctors or the characteristics of individual women, but with the societal effects and implications of the enmeshing of medicine/health care in the wider society. It is these effects and implications that we wish to try and unravel in this book.

The Geographical Perspective

It may seem strange to give pride of place as the final perspective examined to geography, especially as we have already examined part of this approach in our discussion of the behavioural perspective. It is also true to say that the geographical approach could be subsumed under all the previous headings. We wish, however, to isolate the geographical, because it is our own starting-point. In fact, medical and health care geographers have identified several approaches to their subject.

In medical geography, Pyle (1977) has related the specific approaches – medical ecological, medical cartographic, ecological associative, disease diffusion, modelling-simulation and behavioural – with different spatial scales – international, national, regional, interurban, intraurban, household and individual. This scheme is taken by Phillips (1981) as the basis for his own discussion of the geographical perspective. First, Phillips identifies medical cartography in which maps are utilised to depict disease distribution and diffusion as well as areas of deficiency in service provision. Secondly, medical ecology and ecological-associative studies are based on the relationship between disease and environment. Such studies form the basis of part of Chapter 3. Thirdly, diffusion studies plot the spread of diseases often using mathematical models of diffusion. Thus, Haggett (1976) looked at the many directional diffusion of measles using a hybrid version of the epidemic model. Fourthly, modelling and simulation exercises often focus on the location and use of services. Thus the gravity model adaptations to simulate patient to facility flows are included (see Morrill and Earickson, 1969b and Chapters 4 and 5). Fifthly, there are social areas studies which attempt to link disease and health care patterns to the environment conditions of cities. We see such studies as being part of the ecological tradition so important in geographical work and discuss some examples of them in Chapter 3. Sixthly, Phillips isolates behavioural approaches, with their emphases on the individual as the unit of analysis. As previously stated, we regard the behavioural perspective as an essential but limited corrective to the economic and modelling approaches (see Chapter 5). Finally, as an alternative explanation, Phillips identifies the political economy approach, which emphasises the importance of the social, economic and political context within which ill health and health care occur. We regard this as the most important approach identified by Phillips. Indeed, it is the approach which informs much of our work, most explicitly in Chapters 2 and 6, although it is the basis of our critical appraisals in other chapters.

Finally, we may point to Giggs' (1979) attempt to integrate geographical approaches. He identifies three elements: the spatial patterning of illness; the spatial patterning of the physical and human environmental characteristics which adversely affect man's state of health; and the spatial patterning and use of health care delivery systems developed to fight the diseases and hazards which affect health. Such an integrative approach is important as it conjoins geographical endeavour under three headings. It still, however, concentrates essentially on spatial outcomes. As with Phillips' more detailed scheme, there is emphasis on environment, ecology and accessibility, the traditional concerns of human geography. It is not possible to deny the significance of such factors. Indeed, much of this book consists of an exposition of such phenomena.

But, as we have explained, the geographical perspective is merely our starting point. Variations in the distribution of disease and treatment, in the health-enhancing or health-damaging activities groups in space carry out, in the uses made of health care provision and in the allocation of resources between territories are the bases of our approach to health and health care. To explain and understand the causes of such variations we must turn to other perspectives. That statement may not quite capture the spirit of our endeavour in that we do not wish to see one perspective against or added to another. We simply wish to use all available perspectives to help us with our task. Our starting-points may thus be constrained by our training and academic socialisation, we hope that our analyses are not. As social geographers, we see spatial pattern and environmental association as important, but also see the importance of social, economic and political process.

> Human geography itself seems inexorably to be drawn towards even closer association with other social sciences. There is even talk of eventual merger, in some updated version of what used to be political economy. Social geography has played as important part as any subfield in opening up this wider view . . . Social geography's greatest achievement may yet turn out to be the acceleration of its own destruction. (Eyles and Smith, 1978, 55)

Conclusion

We have tried to identify the perspectives that will inform our discussions in the following chapters. We shall not explicitly draw the reader's attention to the relevance or otherwise of a particular perspective in a particular place. We hope that this will be clear enough. We should point to a suggested sequencing for our discussions and analyses of medicine and health care. The book progresses from a questioning of the premisses of medical practices (Chapter 2) and an understanding of the relationship between man, medicine and environment (Chapter 3) through an analysis of health and health care as allocative systems (Chapter 4) and behavioural problems (Chapter 5) to an understanding of the relationship of health and health care in the social fabric (Chapter 6). That sentence gives our sequencing a perfection that is unwarranted. Indeed, to simplify our overall aim is to move from medicine to health to health care to society, or in terms of perspectives from biomedical through geographical, economic and behavioural to materialist. Our central concern is to ensure that 'medicine' and 'health' are firmly embedded in the social fabric. This notion is, therefore, the basis of our final chapter on the directions for health and health care research for 'social geographers'.

2 CHANGING CONCEPTIONS OF HEALTH AND HEALTH CARE IN URBAN SOCIETY

It is a tenet of this book that the ways in which health and health care are seen shape and influence not only the nature of health care provision but also the characteristics of health care research. In this chapter, therefore, we shall be primarily concerned with how conceptions of health have changed over time. We are concerned though not to rehearse yet again the origins, nature and development of the British National Health Service (NHS) but to develop themes discussed in an earlier paper by Eyles (1982). Although the inception of the NHS will enable us to examine different views on the nature of health, we reserve our fuller analysis of the NHS until later, when we discuss the elements of the service to help demonstrate the nature of health care provision. In this discussion of changing conceptions, we shall concentrate on the British experience in particular and that of capitalist societies more generally. Consideration of other experiences – small scale societies, state socialist countries – takes place in the chapter on systems of health care provision. There are then two major sections to this chapter. First, three different conceptions of health will be elaborated. (These are physical health, mental health and social health. It is not suggested that they comprise an evolutionary scheme with each succeeding stage replacing the ones that went before it. Rather the successive stages enrich the conception of health, involving consideration of more and more facets of life.) It will be noted in passing that these conceptions are predicated on certain similar assumptions about the nature of social life and of social relations in urbanised and industrialised society. The second section will challenge these assumptions and examine the inextricable linkages between health, medicine and society.

The Changing Conceptions

In short, it will be argued that there has been a change in the way in which health in urban society is viewed. Thus, in the Victorian city, health was very much a matter of the lack of disease, especially infectious disease, the primary causes of which were seen as overcrowding, insanitation and poverty. In the modern city, health is a dimension of the good life, an aspect of social well-being, while ill-health is seen as deprivation. This change has been brought about not only by the development of national schemes of welfare provision but also by the consideration of mental health and disease. Despite the separation of body and mind in scientific medicine, psychiatric problems are regarded as illnesses that could be treated in the same way as diseases of the body (see Brown, 1973; Doyal, 1979a). Mental health was seen, therefore, as lack of mental 'disease', and equally amenable to scientific analysis and treatment. In the context of urban society, mental health became a dimension of the good life, with the city being initially seen as an environment that produced instability and disorder. Thus in the development of psychiatric treatment in urban society, we see again the change from health as lack of disease to health as a dimension of social well-being. It is possible to draw a subsidiary conclusion – not only does the conception of health change but so does that of the city. Glass (1968) has documented the anti-urban bias of the British ruling class as towns became identified with a working class dissociated from the calming presence of its 'betters'. Further, we can argue that the town ceased to be regarded as the flower of civilisation and became an arena of stress and instability. The links between the urban challenge to individual physical and mental health and its challenge to the social health of the established order can, therefore, be made.

There are, however, three elements of the changing conception of urban health to discuss in detail – health in the Victorian city, mental health and illness in cities and urban deprivation in late-capitalist cities.

The Victorian City and Physical Health

At first sight, it seems unnecessary to trace the origins of the conception of health as the lack of physical disease. The problems of Victorian health are more than adequately rehearsed (see F.B. Smith, 1979). Consideration of such health problems does, however, allow us to raise several important general points. It was in the Victorian era that the debate on the responsibility for an individual's state of health began. The issues of public against private provision and self-help, which still concern health research today, owe much to the conceptions of health and realities of health practice found in the nineteenth century. Further, the emphasis placed on environment as a causal factor resulting in poor health states can be found in much present-day medical geography. Indeed, the emphasis on ecological association and on the individual as a target for environmental factors in medical geography have direct antecedents in the Victorian concern for physical health. We shall return to consider such emphasis more fully, especially in Chapter 3.

The Victorian era was one of high birth rates, declining death rates and rapid population growth. But not only did the population grow, it also became concentrated in industrial towns. Many health problems were created by the poor sanitation and overcrowding in working class areas. But these problems were not necessarily confined to such areas. Ashworth (1954, 47) comments: 'every man of property was affected by the multiplication of thieves, everyone who valued his life felt it desirable not to have a mass of carriers of virulent diseases too close at hand'. In Glasgow in 1840 houses were:

> unfit for sties . . . there is scarcely any ventilation, dunghills lie in the vicinity of dwellings; and from the extremely defective sewerage, filth of every kind constantly accumulates. In these horrid dens the most abandoned characters of the city are collected and from thence they nightly issue to disseminate disease and to pour upon the town every species of crime and abomination. (Quoted in Howe, 1972)

In his 1842 report, Chadwick (1965) noted ill-constructed houses, often mud-walled with ill-ventilated rooms and the retention of refuse inside the house in cesspools and privies. Closeby were dunghills and open drains. Also present were open slaughterhouses, polluted water and filthy streets. The diet of the residents of such areas was also bad, consisting mainly of tea, potatoes and alum-whitened bread. Much food – tea, beer, mustard – was adulterated. Milk supplies were heavily infected with tuberculosis.

Such conditions resulted in many epidemics, such as typhus – the product of working class living conditions – up to 1847–8, and cholera – the result of polluted water supplies and poor drainage – up to 1866. It was not, however, the conditions of the slums that prompted government action, but the threat to social stability that such conditions might well engender. The early reforms in fact did little to improve the lot of the working class. The Poor Law legislation of 1834 made it more difficult for the poor to obtain relief while the Public Health Act of 1848 merely empowered local boards to appoint medical officers of health. The fear of revolt – enlightened self-interest – thus led to change, although we should not underestimate the importance of Christian altruism among social reformers. It may be possible to see interest and altruism acting together in the Poor Law legislation of 1834. As Collard (1978) points out feelings of altruism may well be high when the personal marginal opportunity cost is low. We might further add that self-interest may be served by the very selectivity of such legislation. Selectivity means the creation of scarcity and the need for rationing. The application of strict procedures for the allocation of rational resources is an effective means of social control. While we can identify in the Poor Law legislation the importance of means-testing, the principle of less eligibility and the role of welfare as a force for social control, it would be churlish to suggest that all altruism was so motivated. Many contemporary social services stem from the activities of religious groups. The voluntary sector in particular – hospitals, housing, social work – owed much to the ideas about Christian responsibilities and duties and the nature of Christian fellowship. But as Tawney (1966) points out, there is no clear-cut relationship between religious belief and the care of the poor and the sick. Religion may criticise inequality or regard it as serving some useful function (see Chapter 6). Nor were philanthropists merely concerned with the plight of the poor and the sick. Thus Peabody's bequest of about £500,000 to construct healthy dwellings mainly in London also produced an adequate rate of return (see Tarn, 1966).

Poverty and ill-health were seen as being inextricably linked but not, as Doyal (1979b) points out, in the obvious way of poverty causing ill-health but by the belief that disease produced poverty amongst its victims and, therefore, added to the escalatory cost of poor relief. Economic rationality became the most important justification for public health legislation. The 1850s and 1860s saw the passing of sanitary acts covering the larger towns. But, as Ashworth (1954) points out, the social reformer was regarded as a spendthrift. He quotes an 1858 document which argued that the reformer's 'political science is the science of benefitting the metropolis at the expense of the general taxation of the country . . .

it is not the science of good, cheap and honest government'. Reform, poor relief and sanitary schemes had to be justified in cost-effective terms, by economically rational criteria related to systemic imperatives.

The causes of poverty and ill-health and the limitations of reform were not seen in societal terms, except by Engels (1969) who saw poverty and disease as direct consequences of the way in which the economy operated. The limits to reform were also set by the sanctity of private property which was to remain inviolate. Such limits were maintained in part by pathological explanations of poverty and ill-health, which were seen as the results of individual failure rather than social conditions. Mendicity had to be repressed. A distinction was drawn between the deserving and undeserving poor, with only the former being aided. And, even then, self-help was the mainspring of relief, once the insanitary conditions were removed. Urban health measures, therefore, provided the bases for individual improvement. Indeed, too much assistance would encourage people not to work and would pauperise and dehumanise the poor. Personal and moral guidance were just as important as material assistance. The 1860s and 1870s saw, in fact, the beginnings of modern social work, particularly through the Charity Organisation Society (COS). As Woodrofe (1962) has said, for the COS 'true charity, administered according to certain principles could encourage independence, strengthen character and help preserve the family as the fundamental unit in society'. The members of COS firmly believed that men should be left free to better themselves unhindered by other men and the state. Support might be required temporarily and could be given under the guidance of a COS voluntary worker. The COS clung to this social philosophy despite the emergence of more radical ideas in the 1880s – the reformist ideas put forward by Barnett and Booth in particular (see Stedman Jones, 1971). Such ideas gained support from the surveys of Booth and Rowntree which shook Victorian complacency (Rodgers, 1968; Fried and Elman, 1969). In York, Rowntree (1901) explicitly revealed the link between poverty and ill-health as did the high rejection rates among army recruits and the continuing presence of malnutrition (see Roberts, 1971; Howe, 1972). Such research, especially that of Booth, was mainly concerned with the attributes of the poor rather than the structural mechanisms identified by Engels.

While there was recognition that a large proportion of the

population was poor and unhealthy, disagreement remains as to the level of improvement in working-class living conditions in the nineteenth century. Hartwell (1971) suggests that there was a continuing rise in real wages while Hobsbawn (1969) argues that there was no significant general improvement. While much depends on the dates of the period under investigation, improvements in real wages appeared to have been lower than the rise in *per capita* real income. From the mid-1840s, too, there was a significant growth in real incomes, brought about in the main by falling prices (see Saul, 1969). As Hobsbawn (1969, 162) comments, 'the most rapid general improvement in the conditions of life in the nineteenth-century worker took place in the years 1880–95'. Such improvement resulted, then, from national economic fluctuations and from environmental changes in sanitation and housing provision. This change was but one element in the transition being made from an emphasis on civil rights to one on social rights. As Marshall (1965) argues, early capitalism extended to all the civil rights to dispose freely of property and labour in the market. Such civil rights are the basis of formal equality but in capitalist society they result in great inequalities based as they are on initial endowment of property and skill. Civil rights guarantee the individual non-interference but, as Titmuss (1968) comments, they allow the social costs and diswelfares of the economic system to lie where they fall. Social rights are, on the other hand, legally guaranteed. They are publicly defined and guaranteed claims to certain life chances, particularly those distributed through the health and welfare services. Such a transition does not, of course, simply happen. It may result from the moral obligation felt by the state towards its citizens or from the desire to deflect and contain social antagonisms by providing 'outcasts' with a greater share of national resources (see below). In any event, the impact is to place in centre stage the issues of inequality, poverty and ill-health.

From this brief and partial survey of physical health in Victorian times, two related conclusions can be made. First, the emphasis on the attributes of the poor and the awareness of the large numbers of people in poverty resulted in the gradual rejection of the idea of individual demoralisation, although pathological explanation remained dormant to re-emerge later. Stedman Jones (1971) argues that the necessity of separating the respectful but poor working class from the residuum to obtain broader support for the existing order of things led to the notion of urban degeneration as the cause of urban poverty. This 'switched the focus of enquiry from the moral inadequacies of the individual to the deleterious influences of the urban environment'. As we shall see, this 'deleterious influence' was seen to affect mental health and later the entire quality of urban life. Secondly, the large amount of poverty and sickness led to pressures for reform – public health acts, social security provisions – to enhance the life-chances of the working class but to contain their aspirations within a modified capitalist framework. (Both the enhancement and containment were greatly assisted by increases in productivity and prosperity which are beyond the scope of this book.) As Kincaid (1973, 48) argues with respect to the Beveridge plan – simply a later and more fully developed version of the reforms of the Victorian and Edwardian eras:

> in his plan Beveridge gave expression to a broad section of hardheaded opinion in the ruling class of his period. He was for social reform, so long as the existing structure of society remained fundamentally the same. In fact he was for social reform precisely *because* it would allow the existing order to continue essentially unchanged. It is not as a visionary that Beveridge deserves to be remembered. Rather his particular distinction lay in an ability to translate the general objectives of ruling class reformism into detailed and technically workable proposals.

The emergence of welfarism – of social rights – is one element in the transition from liberal to late capitalism with the state carrying out many economic systemic imperatives. The increased production of 'health' and the diminution of the effects of poverty – 'the improvement of the material infrastructure' (Habermas, 1976, 35) – help to avoid instabilities, by the state replacing the market mechanism to improve conditions for the realisation of capital. We shall return to these issues in the second part of the chapter after looking at urban degeneration and mental health for in these there appear to be shifts away from 'health' as individual freedom from disease to 'health' as this freedom *and* the avoidance of certain environmental conditions.

The Urban Environment and Mental Health

While we shall have more to say on the geographical treatment of mental health in Chapter 3, we feel that the various conceptions of such health help to isolate important ideas and assumptions.

The theory of urban degeneration may be seen as part of the general anti-urban nature of British social thought, as indeed may the idea of urban mental health. While we focus on urban social policies rather than the practice of individual doctors and psychiatrists, the view that mental health, as the product of specific milieux, contributes to the definition of health as a dimension of social well-being must be addressed. Since the Industrial Revolution, the town has been compared unfavourably as a place to live with the village and rural areas. With the concentration and growth of population in towns, the old way of life appeared to be passing away and with it the stability and continuity of social life. The stability and orderly social hierarchy in the village helped to guarantee traditional elite power and privileges. Localism provided the basis for the translation of potentially unstable, coercive relationships into those of stability, order and harmony by reducing the access of subordinate groups to alternative definitions of their situation and by ensuring that elite evaluations were shared moral judgements (see Bell and Newby, 1976). Stability and continuity are, therefore, ideological glosses on elite monopoly of power within the locality. Thus, the move to towns and the break-up of the rural way of life would result, for the elite at least, in a loss of peace of mind. This loss became generalised and was seen as affecting all individuals who entered the anonymous town. The loss of stability, of peace of mind, affected all in the move from rural-based civilisation to urban chaos. This view finds its best expression in the work of Tönnies for whom Gemeinschaft, and its ecological derivative the village, exemplified the settled way of doing (and thinking about) things (see Jones and Eyles, 1977). Because of their ideological significance, the power of such ideas about rural and urban life persisted. As Williams (1973) has said:

> it is a critical fact that in and through these transforming experiences English attitudes to the country, and to the ideas of rural life persisted with extraordinary power, so that even after society was predominantly urban its literature, for a generation, was still predominantly rural; and even in the twentieth century in an urban and industrial land, forms of the older ideas and experiences still remarkably persist.

And speaking specifically of Victorian England, Young (1936, 21–2) commented:

In correspondence with its traditional structure, the traditional culture and morality of England were based on the patriarchical village family of all degrees: the father worked, the mother saw to the house, the food and the clothes: from the parents the children learnt the crafts and industries necessary for their livelihood, and on Sundays they went together, great and small to worship in the village church. To this picture English sentiment clung. It inspired our poetry; it controlled our art; for long it obstructed, perhaps it still obstructs, the formation of a true philosophy of urban life.

Urban industrial life was seen, therefore, as leading to a greater dependence of individuals on others (the increasing division of labour), a loosening of the traditional social controls and an increase in the tempo of life. These features affected people adversely. For example, Simmel was directly concerned with the effect of urban life on mental life, as he called it (Woolf, 1950, 410):

With each crossing of the street, with the tempo and multiplicity of economic, occupational and social life, the city sets up a deep constraint with small town and rural life with reference to the sensory foundations of psychic life. The metropolis exacts from man as a discriminating creature a different amount of consciousness than does rural life. Here the rhythm of life and sensory mental imagery flow more slowly, more habitually and more evenly.

Thus, to deal with the threatening, uprooting currents of urban life, man uses his head rather than his heart. He becomes indifferent, calculatory, blasé, individualistic. But as an individual he is 'a mere cog in an enormous organization of things and powers which tear from his hands all progress, spirituality and values in order to transform them from their subjective form into the form of purely objective life' (Woolf, 1950, 422). In a way, Simmel seems to be saying that man becomes reified, alienated from his true self which is located in traditional society. He is without spirit or true mentality. Broadly defined, mental health has thus deteriorated with urbanisation and industrialisation: a view close to Marx on alienated man. This conceptualisation of the city as alienating is taken further by Smith (1980) who argues that theorists such as Simmel, Freud and Wirth mislead in focusing attention on the city as the cause of social ills rather than on the changing pattern of capitalist economic development.

Indeed Simmel's view on mental life is supported by Wirth, Park and the other ecologists. It is this connection between ecology and mental health and illness that provided medical geography with many of its insights, not necessarily at the level of causation but at that of association (see Chapter 3). Of the ecologists, Wirth (1938) was concerned to show what forms of social action, organisation and life emerged in cities. As cities grew in size, it was less likely that any resident would know all others personally. This meant that social contacts are likely to become impersonal, transitory and segmented with social relationships being treated as means to individual ends. Thus the urban resident loses some of the spontaneous self-expression, morale and sense of participation that comes from living in an integrated society. Morris (1965) suggests that by integrated society Wirth presumably meant traditional rural life. Park (1967) saw the city as possessing a moral as well as a physical organisation. This moral order can affect the mental health of *all* urban residents.

The city thus imposes a particular mental life on all its inhabitants – the *normal* way of life in the city. Not all inhabitants are similarly placed to cope with these demands and pressures. Pathological explanations re-emerge with respect to mental disease and disorder. It is therefore in the attributes of specific individuals and areas that we find the reasons for such 'problems'. Urban degeneration provides the potentiality, the predisposition towards mental ill-health. The nature of individuals themselves – their inabilities to cope with the competition, calculation and rationality of everyday urban life – themes that will reappear below – provides the actual causal mechanism, which architects and planners have attempted to influence at different times by the differential arrangement of space – for example neighbourhood units, new towns, which were meant to overcome mental stress and enhance personal integration.

The ecologists themselves were specifically interested in localities. Faris and Dunham (1939), for example, found that cases of mental disorders, as plotted by residences of patients previous to admission to public and private hospitals, showed a regular decrease from the centre to the periphery of the city. They also found that each of the main mental disorders had characteristic distributions (see Chapter 3). In his introduction to their book,

Burgess (1939) noted that these findings indicate a striking relationship between community life and mental life. 'It may then turn out that human beings in a large city tend to be segregated according to personality types', apparently falling into the ecological trap of imputing areal characteristics from individual data. Dunham (1937) himself was more circumspect, seeing schizophrenia concentrated in the underprivileged parts of the city but manic depression more related to personality and psychological factors rather than the urban environment. As we shall see, many medical geographers appear to support Burgess. The environment is seen as a key phenomenon, while the individual sufferer, if considered at all, is seen as a type with few social attributes and with a greater or lesser predisposition to attract alienating forces. These forces simply exist. They appear to be created by neither the individual nor his social context. They are available and he may or may not 'associate' with the environment that contains them. But as Gudgin (1975) suggests some environments are 'containers' because schizophrenics, for example, drift from random locations into areas of poor housing and little competition.

Psychiatric disorder is therefore associated with a variety of factors: for example low self-esteem (Brown and Harris, 1978), family relationships (Wolff, 1973) and class (Brown *et al.*, 1975). While there has been a shift away from simply seeing particular environments as being alienating and disorienting to a consideration of categorical social attributes – age, sex, occupational status – there is an emphasis on the characteristics of individuals as leading to a greater or lesser propensity towards mental ill-health. The individual is not seen as part of the social fabric, in that if he or she was not present then the social fabric would not be greatly different. In other words, there is no dialectical relationship envisaged between individual and society. Indeed it is possible to see that mental 'disease' is viewed in exactly the same way as physical disease. The normal functioning of an individual organism is disturbed with adverse effects. Individual attributes 'explain' why the organism is functioning 'abnormally' with the broader social system taken as given and, therefore, as unproblematic. Thus the context of disorder is seen in terms of the inability of the organism (person) to cope with 'problems' – an individualistic explanation to which again we shall return. But why does a particular mental state constitute a problem? In the ecology of schizophrenia, for example, individuals who drift into rooming house districts often fail to cope with the demands placed on them by the labour and housing markets. In other words, they cannot cope with, among other things, the demands of an acquisitive society. To judge an individual performance as abnormal also demands some standard of normality. Any definition of normality is of course a social construct. Thus Laing (1960) has argued that schizophrenia is a theory not a fact, the product of a label applied by a group of doctors. Szasz (1970) argues that mental illness is a myth, a mechanism for social control. Normality becomes defined in a particular way with adverse consequences for those seen as abnormal. Indeed Kennedy (1980) argues that we should look at mental illness only in the context of mental health and normality and in that we shall be more aware of the problems of deciding whether a particular state requires medical treatment to return it to 'normality'. If the definition of mental illness is that which is different, it becomes only too easy to label certain forms of behaviour as mentally abnormal. The Soviet Union is an example of a society that uses psychiatry as a form of behavioural control. So, too, did Nazi Germany. Indeed Güse and Schmake (1980) trace the development of psychiatry in Germany: psychiatrists were part of the imperialist–capitalist class and had in the post-Bismarck new economic order the function of regulating the insane and social misfits. The adoption of the techniques and method of science led psychiatrists to an exclusive concern with apparently measurable factors – such as genetically determined ones – rather than the more qualitative impact of social influence and context. Psychiatry (and its science) therefore became the instrument used to legitimate an authoritarian state's drive to isolate those who threatened to disrupt society or were economically unproductive. Their practice thus helped to strengthen the racial hygiene ideology of the Third Reich. While we shall take up the issue of the role of science below, we may comment that concentration on individual functioning is consonant with a specific definition of normality. Indeed the definition has such specificity, that questions of what constitutes normality are seldom asked.

At first sight looking at mental health and cities appear tangential to the chapter. It is rather an extension of the argument. 'Mental health' was challenged by the concentration of population in cities. Normal life was disrupted with concomitant problems that are worsened by the ideological functions served by a settled 'normal life'. City life became seen as normal, though it contained within it the possibility of mental dislocation. Certain individuals and groups – often those unable to adapt 'satisfactorily' to the rational, calculating way of life that dominates late capitalism – are seen as having a greater propensity towards mental disorder. They can be treated as individuals who cannot function normally in urban and individual life. The definition of this life as normal remains largely unchallenged, because the majority manage to cope albeit in an alienated way. It is not until we begin to challenge the definition of normality that we get below the surface of everyday life and discover the structures that shape our lives (see below). This is not to deny that disorders of a biophysical and biochemical nature exist, but peace of mind, quality of life, depend on a specific definition of normality. And this definition relates more to systemic imperatives than personal ones.

Urban Deprivation and Social Health

'Health' has, therefore, changed. Initially seen as physical health (or lack of physical disease), it became broadened at the end of the nineteenth century and the beginning of the twentieth century to include mental health or normal functioning in an urban and industrial environment. It has since been broadened further, particularly in the post–1945 era and from the impetus of reformist ideas. Indeed, the Second World War, along with the Depression, provided the catalysts bringing together many of the reformist themes of the first third of the twentieth century. Marshall (1975) points to the many wartime statements concerning the creation of a comprehensive hospital service, equality of opportunity in education and the implementation of full social security. The welfare state was, in fact, created by the welding together of measures of social policy into a whole, with a safety net of national assistance to help the unemployed, the old and the sick. Welfarism helped redefine 'health' – it became the social 'health' of individuals and families not merely the lack of physical and mental disorders. With the redefinition of health, we see, too, the transition from civil to social rights.

The granting of social rights by the state has not meant the application of uniform principles of allocation. Different welfare agencies apply different principles, suggesting that the practices of the state can often be contradictory. Different agencies may have different conceptions of need and of their functions in serving that need. Indeed bureaucracies may well take on values and goals of their own which may conflict with the aims of other state institutions (see Chapter 6). It is possible for an institution to regard the individual who is poor or sick as requiring minimum assistance to enable them to help and provide for themselves. Individual responsibility is promoted, and financial resources are provided to meet basic human needs. Such benefits are vigorously means-tested, that is they are provided selectively, such as, for example, discretionary payments to the unemployed and the long-term sick. Those applying for such assistance are stigmatised, this as Jordan (1974) argues, being intended to deflect the resentment of the subsidised by the lowest wage-earners and thus reduce potential conflict. Assistance is, therefore, a form of social and, indeed, behavioural control. For example, the personal social services, one of whose functions is to help militate against the alienating features of modern urban society, may withhold discretionary payments as a way of sanctioning the social behaviour of their clients. Thus the clients' own responses to their individual inadequacies perceived by social workers are important in determining their opportunities to receive benefit. We see again the features present in the nineteenth-century assessment of poverty and ill-health, namely individual inadequacy and responsibility, stigmatisation and selectivity. Individuals are divorced from their social contexts and seen as not only behaviourally irresponsible but also morally inadequate. This notion of moral adequacy will reappear below; but we can see that the way to good health is seen through behavioural modification and the selective aid of the state is geared towards the moral re-education of the inadequate to make them normal functioning entities once again.

Alternatively, assistance may be provided according to universalist principles – to each according to their need – which, overtly at least, foster social integration rather than effect social control. In the application of such principles, there is recognition that the collective provision and universal enjoyment of such services as the NHS must take precedence over individual rights. The acceptance of universalism requires individuals and groups to acknowledge that they are members of a common moral community. In Marshall's (1965) view, such common status is based on the social right of citizenship. As Pinker (1971, 100) notes, universalism rests 'its moral claim on the ethics of co-operation and mutual aid'. It rests, therefore, on the collective consciences of citizens to provide adequately for collective needs. This belief in the essential altruism of individuals seems somewhat idealistic in our apparently competitive and acquisitive world. It did, however, underpin Hall's (1957) conception of welfare and health care. She argued that social service was essentially the manifestation of a personal interest in the human condition and that the welfare state is a witness to the reality of our belief in community. But, as Baker (1979) argues, such a conception suggests that social policy manifests, through the state, the love that people have for each other and that changes in such policy are cumulative, irreversible and result from a widening and deepening sense of obligation. Such a view sees the rise of the welfare state as a romantic enterprise (see Hay, 1975). In other words, the notion of integration through citizenship is somewhat problematic. It assumes an equivalence of social rights that is not present despite the formal equality granted by citizenship. Indeed, Marshall (1965) himself was not certain that such a commitment to universalism would emerge. Further, it is possible to argue that such a conception of health care is asocial, seeing individuals as pre-social phenomena consisting of bundles of equivalent rights. The problem of assisting the sick and needy becomes a simple matter of fine-tuning, of tinkering with administrative and bureaucratic machinery to ensure a more equitable distribution of resources (see below, and Chapter 6). This view greatly underemphasises the political and economic constraints on policy-making. In fact, Wedderburn (1965) argues that such constraints mean that only a residue is left of the universalistic framework of the welfare state, a residue that tempers the prevalent values of competition and acquisitiveness. Such tempering serves important societal functions. Health and welfare systems are adaptive mechanisms for social survival and tension management. According to Titmuss (1968, 100) social welfare is 'a major force in denying the prediction that capitalism would collapse into anarchy'. While such functionalistic explanation is not without difficulties, it does demonstrate that a consensual, integrative view of health and health care has societal implications. Indeed, following O'Connor (1973) we would go further and suggest that the collective expenditure on social policies, that is the appeals to universalism, mask the private appropriation of the benefits of such policies. This contradiction is likely to be enhanced by the inadequacy of the state's tax base to finance adequately the improvements in the material infrastructure required for an expansion of the forces of production. Universalism attempts, then, in its appeal to collective interest to minimise material self-interest, and by emphasising individual benefit it reduces consideration of the relative positions of power and advantage of social groupings.

We shall return to these problems below, but it seems obvious from the experience of particular regions and groups (see Chapter 3) that universalism has not secured the equalisation of social rights and health status. Throughout the 1950s it was widely thought that material poverty (and much physical disease) had been largely overcome. (It was not until the 1960s and 1970s that the disease of affluence truly emerged in public consciousness and became part of our definition of 'health'. Indeed, it has become increasingly recognised that the affluent and poor have similar health 'problems' and that their severity of impact is a matter of degree rather than kind.) Rowntree and Lavers (1951) pointed to the declining poverty population of York as well as to the creation of welfare agencies such as the NHS. Such findings though were challenged by Townsend (1954) who argued that the calculations of essential expenditure by Rowntree were not based on the actual spending patterns of working class groups. Nor did he take account of the social context of expenditure which could apply pressure on families. Poverty, health and well-being are therefore relative terms in urban industrial society. They must be defined by the aspirations and expectations of the community in question, in relation to the resources enjoyed by themselves and others. Given the cultural context of late capitalism, where success and the enjoyment of resources are highly valued and where there are mass, virtually instantaneous forms of communication, individuals can sense the deprivation relative to other social members. This is not to say that poverty or ill-health (of a physical or mental nature) have disappeared (see Harrington, 1962; Townsend, 1979) but they are now heightened by their seeing what others have – by a sense of deprivation.

Such feelings were first produced, and may be more intensely felt, in urban areas, where the bulk of the population resides. Besides national welfare and health schemes, urban conditions have been improved by redevelopment and rehabilitation schemes. Modern dwellings replace dilapidated slums. Overcrowding is eased, health improved and more facilities provided, although housing costs rise and old social networks may be destroyed. But despite these many improvements in material conditions over the past 25 to 30 years made possible by general increases in prosperity, a certain proportion of the population remains deprived. It has been argued that these people are concentrated in specific parts of the city, notably the inner city. Thus Castle and Gittus (1957) point to the concentration of 'social defects' in Liverpool, the GLC to areas of housing stress in London and the DoE to specific inner city areas with a coexistence of deprivation in London, Birmingham and Liverpool.

It has been suggested that this concentration is partly due to the age of their housing stock, partly to the state of their economic bases and partly to selective in- and out-migration, with their losing the young and skilled, retaining the old and unskilled and gaining overseas immigrants. These areas have become the recipients of special assistance and initiatives, because of the desire to upgrade them and because of the fear that they could become bases for the development of juvenile delinquency, racial tension and the like (see Eyles, 1979). Educational priority areas, comprehensive community programmes, partnership schemes and enterprise zones have been declared while urban aid expenditure has gone predominantly to inner areas. Such tinkering points, of course, to the apparent inadequacy of universalist procedures. In fact, Titmuss (1968) argues that the choice is not between universal and selective urban and social services but that the challenge lies in obtaining an infrastructure of universalist services to provide a framework of values and opportunities within and around which selective services, aimed at positive discrimination, can be developed. It is possible that positive discrimination may stigmatise, pointing to the recipients as individual failures in need of special, selective treatment. Positive discrimination may also politicise – the group experience often on an areal basis of receiving such treatment may lead to the creation of welfare rights groups or community health councils (see, for example, Topping and Smith, 1977) and a collective critique of the prevailing conceptions of social life and existing allocation procedures (see below).

In any event, most of these declarations have been token demonstrations of possible action, involving minute amounts of public money. Townsend (1976) estimates that the entire urban aid programme constitutes only ¼ per cent of the RSG. These areas of low quality of life (poor social 'health') are in fact not so much the recipients of social policy but arenas of social experimentation. Further, positive discrimination based on ecology misses more of the poor or deprived than it includes. The extent of deprivation (social ill-being, of which physical and mental health are important dimensions with the materially poor more likely to suffer from physical disabilities (see Walters, 1980)) is determined by the national system of resource allocation. The Community Development Projects (1977 a, b) go further, suggesting that improved welfare services only have a marginal effect on deprivation. Poverty and much ill-health result not from 'the pathologies of individuals and areas but from fundamental inequalities in the capitalist politico-economic system, in the very structure of the British economy and polity. Even national schemes such as the NHS may not challenge these fundamental structures. Thus Navarro (1978) argues that the creation of the NHS demonstrates that:

> there is no clear cut dichotomy between the social needs of capital and the social demands of labour. Any given policy can serve both. Indeed, social policies that serve the interests of the working class can be subsequently adopted to benefit the interests of the dominant class.

And it may be that reductions in social and health expenditure represent the fact that, at a certain level, public spending becomes dysfunctional for capital accumulation.

It is now possible to discern some continuities emerging from beneath the changing emphases on different *types* of 'health'. Health is seen as a lack of disease and/or deprivation. This definition is based on a socially constructed state of normality, against which individual functioning and performance are assessed. Indeed, health problems are seen in the main in individualistic terms in that pathological or degenerational explanations are employed to account for 'abnormal' states. Reform (change) also paradoxically illustrates continuity. Limits to reform are set by the existing ways of doing things; indeed by being within those limits, reform can enhance the continuity, capacity and adaptability of the prevailing system by providing a healthy workforce and institutions that can deflect antagonisms over material rewards between capital and labour. Appeals to universalism and citizenship essentially ensure that continuity and adaptability.

These continuities – essentially the reproduction of social life and the social construction of 'health' – form the second major part of the chapter – the context of the first – in which the relations of health and medicine and of rationality and medical practice and the production of medicine will be discussed. The concept of health will no longer be taken as unproblematic, as self-evident. Questions of definition and of context become central.

The Social Construction and Fetishism of Health

Health and Medicine

There is an ambiguous relationship between health and medicine. We tend to assume that health care and medical care amount to much the same thing, but they posit different concepts of man and of society. It is impossible to deny the importance of medicine in alleviating human suffering, especially in the treatment of infectious diseases, the removal of pathogenic organs and the application of chemotherapy, but there seem to be definite limits to the ability of medicine to enhance 'health' or the human condition. Medicine can be conceived as physical or chemical intervention to restore a disordered system to 'normality'. Health, according to WHO, can be viewed, somewhat ideally, as a state of complete physical and mental well-being and not merely absence of disease or infirmity. It is necessary to add a material basis to such a definition, for 'health' has no real meaning without reference to economic and political conditions that contain it.

It becomes possible to argue, by definition, that medicine has had a limited impact on health. As Powles (1973, 12) argues:

> Industrial populations owe their current health standards to a pattern of ecological relationships which serves to reduce their vulnerability from infection and to a lesser extent to the capacities of clinical medicine. Unfortunately, this new way of life, because it is so far removed from that which man has adopted, has produced its own disease burden. These diseases of maladaption are, in many cases, increasing.

Carlson (1975) suggests that the gulf between health and medicine is in fact becoming greater. He cites increases in the complexity, stress and size of organisations and the persistence of work-related stress leading to new aggravated health problems; increases in the diseases of 'civilisation', while medicine maintains an emphasis on cure rather than prevention; an ageing population; a rise in accidental deaths and injuries; continuing environmental degradation ignored by medicine; increases in mental and emotional disorders; an even greater concentration on acute conditions resulting from biomedical advances; the increasing professionalisation of medicine with a reductionist drift to technological solutions and greater depersonalisation; and increasing bureacratisation as medicine becomes more politicised. Thus, along with the other sciences, medicine has elaborated its means and forgotten its ends. Medicine becomes (or became) a realm in which technological solutions are found to functional problems. As Roszak (1972) argues, medicine is based on objective consciousness, on reducing all things to terms that objective consciousness might master. Medicine appears, then, as an independent segment of society run by skilled technicians, or 'experts', the doctors who treat patients as individual organisms and are treated themselves as independent contractors.

Independence and isolation are, therefore, characteristics of modern medicine and its object of analysis and treatment, the body. As Foucault (1973, 189) argues, medicine perceives disease in the individual body, localising causes and suggests that 'the local space of the disease is also, immediately, a causal space'. As we shall see below, medicine is, therefore, the science of the individual. Such an approach is closely related to the idea of the aetiology of diseases in specific pathogenic agents. Medical concepts are further reinforced by the therapeutic practice of isolating the individual from the social context in which the disease is acquired. The disease becomes, therefore, 'professionalised' rather than 'socialised'. In other words, the independence of medicine means that the most important element in explaining and treating the disease becomes the expert – the professional (see Chapter 5) – rather than the environment or social context. This reasoning stems in part from the adoption of scientific procedures in medicine (see below) which help isolate disease from the social fabric. As we shall see in Chapter 6, this view is peculiarly that of scientific, allopathic medicine. In other cultures, illness, medicine and health are not isolated. Nor are they regarded as independent phenomena. Glick (1967, 53) argues, in the context of the Gimi of New Guinea, that 'illness has meaning for the community, not just for an individual' and that treatment is directed to solve conflicts in the social fabric.

In other words, social context and environment become vital factors.

However, this apparent independence of medicine from the social system has led Illich (1977), for example, to argue that the massive increase in medicine has had a largely damaging effect on health. He accuses medicine of expropriating people's health by creating dependency on medical interventions and by removing essential copying capacities of individuals and small local communities. Thus, he argues (1977, 49) that 'medicine undermines health not only through direct aggression against individuals but also through the impact of its social organization on the total milieu'. Social iatrogenesis, the damage to health caused by medicine's sociopolitical mode of transmission, obtains when medical bureaucracy creates ill-health by increasing stress, lowering levels of tolerance to pain and abolishing the right to self-care and turns health care into a standardised product. Elling (1977) and Lall (1977) would go further, suggesting that the specific mode of production – the capitalist mode – results in the establishment of hazardous and polluting industries in specific areas, namely Third World countries, and in export of expensive, sometimes outmoded, drugs. In other words, medicine, and particularly its containing social system, determine patterns of ill-health.

But, in any event, the 'radical monopoly' of medicine and its professionals leads to a technological definition of 'health' with an emphasis on the individual. While correctly demonstrating that health is *not* the absence of medical classifiable disease, Illich, however, fails to recognise that human beings are not totally self-sustaining. His analysis lacks a definition of *necessary* and *justifiable* medical interventions and support services (see also Campbell, 1978). Perhaps more circumspectly, Carlson (1975, v) argues that 'the system of engineering intervention on people and on environments, which constitutes the contemporary medical endeavour and around which the modern medical institution is built, has almost no relevance to health'. He concludes that medicine today sees disease as a result of faulty machinery, susceptible to fine tuning and discrete diagnosis. Such a view is heightened by (and indeed is partly dependent on) the increases in specialisation, surgery and drug flow.

Science and Medical Practice

But should we expect medicine (and therefore 'health care') to be any different? The answer must be no. Powles (1973) sees medicine and medical practice as part of the culture of its containing society, but there is perhaps more to it than that. Medicine is both shaped by and gives support to the existing forms of social order and organisation. Under capitalism, rational–legal authority and rationalistic practices dominate. Indeed, Fay (1975) argues that the centrality of science in social theory and practice generates social engineering solutions to problems. The use of technology is one of the products of the rational, scientific approach to medicine adopted in modern society. Kennedy (1980) points to the avowedly scientific nature of medical education in which medicine is seen as a curative science. Doctors see themselves as engineers, as problem-solvers, dealing with problems as they *arise*. Little attempt is made to *avoid* them. McKeown (1979) further demonstrates that the direction of medical research has been largely determined by the belief that improvement in health depends on knowledge of the body and the application of technological developments to its diseases.

Such a scientific view of medicine and 'health' removes these categories from the experience and capabilities of individuals. Experts must mediate between the individual and his body and mind. This expert is, of course, the physician who can be seen to have enormous power and influence because of this mediating position. 'The medical profession has first claim to the jurisdiction over the label of illness and *anything* to which it may be attached' (Freidson, 1970a, 251). Freidson further argues that the physicians' strength is based largely on their legally supported monopoly over practice. 'While the profession may not everywhere be free to control the *terms* of its work, it is free to control the *context* of its work and the technical instruction of its recruits' (Freidson, 1970b, 84). Thus:

> by and large, within the well-financed division of labour dominated by the profession and under its protective umbrella, most work is limited to that which conforms to the special perspective and substantive style of the profession – a perspective that emphasises the individual over the social environment, the cure rather than the prevention of illness, a preventive medicine rather than . . . 'preventive welfare' – social services and resources that improve the diet, housing, way of life, and motivation of the people without their having to

undertake clinical consultation with a practitioner. (Freidson, 1970b, 148)

The expert and his institutions – especially the hospital – therefore dominate and their very practice helps shape and support broader societal practice, specifically through the individualistic and mechanistic bases of medical care.

Doyal (1979a) argues that the most fundamental characteristic of scientific medicine is indeed its mechanistic approach to human beings. She states that 'in its historical origins, medicine followed the more general pattern of Renaissance science in analysing phenomena as a set of mechanically related parts rather than as an organically integrated whole' (1979a, 238). The doctor's role is to restore the human machine to normal functioning (see McKeown, 1971). The importance of 'normality' reappears while the emphasis on individual malfunctions means that the social and economic causes of ill-health are often neglected. This medical model has parallels in functionalist theories of society which are seen as a set of mechanistically related parts. Social problems are thus caused by the inability of individuals to adapt successfully to the prevailing order. Thus, in the case of body, mind and social order, a particular state is seen as normal, while deviations are seen as problematic abnormalities requiring 'treatment' so that the normal state can be re-established. Indeed, such a perspective is the basis for Parsons' (1951) idea of the sick role, which provides a legitimate basis for the exemption of the sick individual from 'normal' roles. To receive such exemption, though, the individual must recognise that illness is undesirable and seek help. Health is seen as necessary, therefore, for normal functioning in society (see Parsons, 1964). 'Health' can be restored with the assistance of expert professionals. While such a view conflicts with lay conceptions of health and illness and begs questions on the constitution of normality, it does point to the fact that a major social function of medicine is the regulation and control of a specific form of deviance, sickness. That which malfunctions – the body, the health care delivery system itself – must be reintegrated into normal society by the application of technical procedures by recognised professionals. Such a view is, of course, mechanistic predicated on a consensual, steady state view of body and society. Despite its unreality, it remains influential with 'getting well' being widely seen as 'getting back to normal'. Indeed Ehrenreich (1978) argues that the medical profession not only apply technical knowledge but also impart social messages which incorporate a specific ideology representative of dominant groups in society.

The models are not only mechanistic but also individualistic. Stark argues that:

> Disease is understood as a failure in and of the individual, an isolatable 'thing' that attacks the physical machine more or less arbitrarily from the 'outside' preventing it from fulfilling its essential 'responsibilities'. Both bourgeois epidemiology and 'medical ecology' . . . consider 'society' only as a relatively passive medium through which 'germs' pass en route to the individual. (Quoted in Doyal, 1979b, 35)

Doyal goes on to suggest that this individualism has taken a new and more powerful form. The emphasis on 'diseases of affluence' – smoking, diet, lack of exercise problems – means that ill-health can be explained in terms of individual moral failings in which the victim is blamed for what has happened to him (see above). Such an argument is elaborated by Crawford (1980) who sees the self-help and holistic health movements in the United States as having both positive and negative impacts. They reject the medical view of the deficient client. Self-help seeks to reduce the dependency on the professional, while holistic health is a response to alienation in the medical encounter, offering an overtly experiential understanding of disease. Such movements – 'healthism' – modify medical causality towards host resistance and adaptation. Health is seen as a moral state, but in the reclamation of health from medicine, there is a tendency to see individual, moral responsibility as sufficient. The assumption of individual blame is again promoted. Poor health is seen as an individual failing and failures are seen as near pariahs. Indeed, seeing health as moral adequacy is a further medicalisation of everyday life. 'Healthism is a kind of elitist moralising about what are believed to be unhealthy coping behaviours' (Crawford, 1980, 385). Social context is again ignored – individual attributes and pathologies 'explain' ill-health.

Such a view of health:

> strengthens the basic ethical tenets of bourgeois individualism, the ethical construct of capitalism where one has to be free to do whatever one wants, free to buy and sell, to accumulate wealth

and to live in poverty, to work or not, to be healthy or to be sick. Far from being a threat to the power structure, this life-style politics complements and is easily co-optable by the controllers of the system, and it leaves the economic and political structures of our society unchanged. Moreover, the life-style approach serves to channel out of existence any conflicting tendencies against those structures that may arise in society. (Navarro, 1976, 126)

Perhaps we should add especially in medicine where the great ideological power of physicians means that conflicting lay claims can be dismissed by the pronouncement of the 'expert', while lay treatments – herbal remedies for example – may be co-opted by their commercialisation.

This stress on individualism is a central tenet of capitalist society. Habermas (1976, 82) argues: 'Historically, bourgeois society understood itself as an instrumental group that accumulated social wealth only by private wealth . . . Under these conditions, collective goals could be realised only through possessive – individualistic orientations to gain.' The crises tendencies in advanced capitalism have not destroyed the *belief in* or *ideology of* the individual. As Adorno (1974, 251) comments:

In the midst of standardised and administered human units, the individual lives on. He is even placed under protection and gains monopoly value. But he is in truth merely the function of his own uniqueness, a showpiece like the deformed who were stared at with astonishment and mocked by children. Since he no longer leads an independent economic existence, his character falls into contradiction with his objective social role. Precisely for the sake of this contradiction, he is sheltered in a nature preserve, enjoyed in leisurely contemplation.

However atomised and alienated the individual may be in reality, its conception is thus ideologically important, helping to maintain a particular social order. As Habermas (1976, 127–8) says:

The poverty of the bourgeois subject consists, then, in his uncomprehended particularity . . . bourgeois humanism . . . reflects *nothing* more than a decisionistic presumption of a monopoly on the definition of humanity – 'the history of bourgeois society is the history of those who define who man is' . . .

The institutions of the social order – including medicine – thus emphasise the individual. The individual, though, is segmented, parts of his life are seen and treated in isolation, while reference to the social structure is further removed by reliance on pathological explanation.

Medicine and Society

Medicine is thus not simply an institution that enhances health or alleviates suffering, it also has systemic functions, both helping to shape and support a capitalist social order. This order also helps shape and support medicine. Indeed the interactions between capitalist institutions and between these and their supporting ideology are complex. This chapter has attempted, in rather a crude way, to illustrate some of these interactions, namely between health and medicine and between the individualistic, mechanistic models in both medicine and society. In these interactions, medical practice and definition have been seen to be of vital importance. Physicians help to maintain a particular social order. They do this not by any conspiratorial machinations but by the very nature of their ideas and work. We should not overemphasise the role of doctors for they, their institutions and medicine itself are fully embedded in the social order. Thus Navarro (1978, 86–7) has argued that:

> the primary controllers and managers of medicine are not the professionals, but rather the controllers and managers of capital . . . the concept of health and even the nature of medical practice has continuously changed and has been redefined according to the needs of the capitalist modes and relations of production. The medical profession intervenes in that redefinition, but *a posteriori*, i.e. they administer and influence but do not create the nature of medicine.

Such a perspective has some support from Frankenberg (1974) who sees power as the crucial variable in understanding health care and Elling (1981) who argues that health expenditures are only understandable in terms of power relations within the private economy. There may be more to it than this. Physicians appear to have a more active role in shaping medical practice than Navarro accepts. He suggests that the interaction between medicine and the social order is a one-way relationship, but doctors acting as scientific and vocational experts in a specific social context may well instigate changes and reforms that lead to systemic adaptations. Most social practices are not static but react to ideas and activities that challenge their ways of doing. Indeed, institutions created from working-class experience, such as friendly societies, led to significant changes in social practice (see below).

The extent of such reaction may well, however, be limited to that which does not directly impinge on the central elements of medicine and social power. Reaction may be muted by the medicalisation of society which expands rather than restricts the arena of professional competence and control. Zola (1972) argues that as more social phenomena become linked to medicine, doctors act as 'gatekeepers' for more functions. Medicine mediates the assumption and attainment of the good life. Thus lay conceptions and experience – the bases of reaction – confront the logic of 'scientific' procedures and 'reason'. Life-style becomes increasingly subject to the scrutiny, evaluation and control of medicine. Further, medical labelling ensures greater social control of 'problems' (see Conrad and Schneider, 1980). Medicine thus becomes (or remains) an institution of social control.

It is, however, possible to develop the orthodox Marxist analysis of medicine suggested by Navarro. Kelman (1975, 634) distinguishes between functional health and experiential health and orthodox Marxists concentrate on the former which 'is that organismic condition of the population most consistent with, or least disruptive of, the process of capital accumulation'. Health and medicine become ways of reproducing the relations of production. Cockburn (1977, 56) argues that:

> if capitalism is to survive, each succeeding generation of workers must stay in an appropriate relationship to capital: the *relations* of production must be reproduced. Workers must not step outside the relation of the wage, the relation of property, the relation of authority. So 'reproducing capitalist relations' means reproducing the class system of ownership, above all a *frame of mind*.

At one level, therefore, the workforce must be maintained in reasonable 'health', while at another many institutions help shape this frame of mind that sees those same institutions and existing social relations as the 'natural' order of things. This hegemony – organising principle or world-view – is diffused by these institutions into every area of daily life. In fact, Krause (1977) argues that such hegemony is powerful enough, in the case of Mexico at least, to change attitudes to the traditional systems of care. Through socialisation, however, such hegemony becomes internalised and part of commonsense, the generalised way of looking at the world. Thus medicine is seen as health and this view helps preserve the established medical order and reproduce capitalist relations of production.

It is possible to examine the appearance of medicine as health in terms of the production of commodities. Medical care is sometimes seen as a saleable commodity (Carlson, 1975; Rhodes, 1976). Illich (1977, 50) hints at the fetishised nature of modern medicine when he argues that radical monopolies 'impose a society-wide substitution of commodities for use-values by reshaping the milieu and by "appropriating" those of its general characteristics which have enabled people so far to cope on their own'. The output of non-marketable use-values (e.g. self-care, mutual aid) is also paralysed. Illich does not, however, extend his assertion into analysis or locate the need for such medical care in the reproduction of the relations of production.

Medicine can be analysed as a commodity in the Marxian sense. Marx (1970, 35) argues that 'a commodity is, in the first place, an object outside us'. Avineri (1968) in his Hegelian-influenced interpretation of Marx argues that 'in the first place' implies that a commodity may ultimately be something else: indeed ultimately an objectified expression of an intersubjective relationship. Thus a commodity relationship displaces or disguises a social relationship. In the medical world, it becomes a technological, mechanistic relationship in which the doctor – the independent contractor – repairs an object, the human machine. This inversion of subjective into objective relationships is, of course, the fetishism of commodities. Human activity and relationships appear as natural objects (Marx and Engels, 1970). People become objects and objects receive human attributes. Human relationships become transformed into exchange relationships in which all participants are stripped of human characteristics. Thus:

a commodity is therefore a mysterious thing, simply because in it the social character of men's labour appears to them as an objective character stamped upon the product of the labour – because the relation of the producers to the sum total of their own labour is presented to them as a social relation, existing not between them, but between the products of their labour. This is the reason why the products of labour become commodities, social things whose qualities are at the same time perceptible and imperceptible by the senses . . . There is a definite social relation between men that assumes, in their eyes, the fantastic form of a relation between things . . . This I call the Fetishism . . . which is inseparable from the production of commodities . . . . (Marx, 1970, 72)

The social bond between individuals thus appears as an exchange-value (the relative market power of a thing *vis-à-vis* all others), values which become more and more dominant in bourgeois society (see Marx, 1973). Capitalism universalises the scope of market and exchange relations and gives rise to the most extreme form of alienation, that of man from his own species-being.

In essence, therefore, medical relationships are exchange relationships, in which market power is based not solely on labour market position but also on those Weberian additions to market situation – skill, knowledge, status. Further, medical relationships are objectified relationships in which the doctor as engineer, technician, rational scientist and independent and quasi-independent contractor confronts the patient as depersonalised consumer, as malfunctioning machine. Such analysis extends the arena of commodity production and fetishism away from work to social life in general. In fact, some are critical of the notion of commodity fetishism and its extension (see Leiss, 1978). But it is an extraordinarily rich concept for not only does a commodity have use-value in its satisfaction of specific needs and wants, but also under particular historical circumstances, its exchange-value – its form – becomes dominant. It is under such circumstances, under capitalism, that the commodity takes on a life of its own – that it leads to the reification of social relations and the reproduction of particular social forms. In our limited context, therefore, we can say that medicine is a form of commodity production and exchange and represents the fetishised nature of health in capitalist society. In this way, it helps reproduce itself as a form of activity and as an institution and certain relations of production.

Hirsch (1977) has introduced a modified version of commodity fetishism which is also relevant to our conceptualisation of health and medicine. He argues that once basic material necessities are met for the majority in a market-exchange economy, there is intensified competition for positional goods, which are by their nature scarce and are used to define social status differences between individuals. Because of intense competition for them, benefits are small while costs are high. The positional economy is in fact largely an expression of a bias towards material commodities and of a commercialisation effect. Thus, individuals increasingly interpret their well-being as the degree of success they have in changing their possessions and life-styles – a form of commodity fetishism. It is possible to argue that the search for a healthy life-style is part of this fetishism – diet, exercise, medical insurance all play a part. This Hirschian notion of commodity fetishism fits well with a systemic imperative look at earlier-individualism, especially that modern version which indicates that individuals are themselves responsible for generating a healthy environment, irrespective of any structural constraints that may be operating. Commodity fetishism, individualism and pathological explanation combine to help reproduce relations of production.

This second section of the chapter has tried to look at the societal or systemic context of the changing conceptualisation of 'health' and 'health care'. It suggests that beneath the change from physical health to social well-being, there lie societal needs – primarily the reproduction of the relations of production – that *in part* dictate the change. It was also suggested that medicine and medical care are seen as being synonymous with health and health care for systemic reasons. Thus, in the first part of the chapter we are in reality looking at changing medical interventions, that change being brought about by the tensions between the needs for system maintenance and the pressures for reform. Therefore, while medicine is fetishised 'health' and takes a commodity form, change does occur, although it may be argued that it only takes place within certain limits, limits that challenge neither the basic relations of production nor the basic structure of medical practice. Indeed, it has been suggested that medicine and the social order shape and maintain one another as part of the totality of capitalism.

Change and Reform

While this chapter has been concerned with looking at changing conceptions and at the changing nature of medical practice and intervention, little has been said about how such changes have been brought about. This is not the place to discuss the nature of change in all its entirety, but while reference will be made to specific changes and reforms in Chapter 6, some attention must be given here to such problems. One view is to see science and medicine as evolutionary, rising to meet the challenge of technological and social improvements. Thus increased car ownership and greater prosperity bring the problems of road accidents and obesity. Change is therefore unproblematic: its hand-maiden, progress, will eventually solve the problems.

It is possible to see change and reform as dependent on the distribution of power and advantage. It is then the collective experience of a deprived group that demands change, a concomitant of which will be a different view of the world. Thus, on the development of the NHS, Navarro (1978, 11) asserts that:

> there emerges a popular demand for assuring the availability of services . . . And that demand eventually demands a response from the dominant class, a response based on the necessity for that class to legitimate the social order in which it holds dominance. And that dominance further reflects itself in the nature of the response, i.e. it primarily mirrors and reproduces the class interests and ideology of that dominant class. Indeed . . . the specific interest groups, including, among others, professional interests, creep in and substantially shape its form.

Such a perspective does scant justice to the activists and reformists in the medical profession themselves. Waitzkin (1981) points, for example, to the importance of Virochow and Allende in helping institute changes in health conception and practice. Further, as Urry (1981) points out, Navarro really fails to show the connections between industrial militancy and health-insurance legislation 20 years later. He only asserts that the legislation stemmed from the demands of labour and the legitimising needs of capital. Not only does he see the state as only functioning to legitimise (rather than also to organise the appropriate conditions for capitalist continuation), he also fails to consider the nature of power blocs and alliances. In other words, experience and demand operate in rather a mechanistic way to produce a reform.

We may be better served by following Kelman's (1975) distinction between functional and experiential health. The latter grows out of the relations of production, while the former is largely elicited from the forces of production. (Thus we can see why so many institutions direct their endeavour to the reproduction of the *relations* of production.)

> Experiential health is simply people's own conception of what it is to be 'healthy', and this conception may vary widely with respect to objective accuracy . . . In capitalist society, with continued conflicts between the accumulation process and the process of social development, there are two inherent contradictions in the definition of health: that between functional and experiential health and that within the social determination of experiential health itself . . . 'health' in a capitalist society is nothing more than the prevailing standoff at a point in time between its functional and experiential aspects, between the tendency for the accumulation process to reduce its subsumed human populations to the status of resources employed for its expansion and the tendency of people to seek their own transcendent (of the accumulation process) fulfillment. (Kelman, 1975, 635–6)

Thus, Kelman is suggesting that change comes from 'below', from a group's own definition of its health state in relation to that of others and objective conditions. We shall add that such reform (change) may be spontaneous but it is more likely to take the form of organised, reformist groupings that may well be led by doctors and politicians.

While lay concepts will be discussed more fully in Chapter 7, reference should be made to Smith's (1981) study of black lung which demonstrates the connections between changing definitions, relative power and advantage and working-class experience. She points to three major definitions of this mining disease. First, and related to initial mining operations in the anthracite fields of Pennsylvania, black lung was a matter of speculation, based on observations of the unusual respiratory disease burden of miners. This initial stage was soon overtaken by the second, illustrated by the expanding bituminous coalfields of South Appalachia and in which there was a stark class structure and tight corporate control of the health care system. These company doctors were the only source of medical care. As disease was an economic liability to the companies, diseases were attributed to the fault of the miner. Respiratory trouble was often seen as malingering or compensationitis, a stigmatised sign of psychological weakness or duplicity. Such a view supported the class structure of the coal camp, which gained further from the prevailing 'germ' theory of disease causation which implicitly denied a role for social and economic factors. This system did not, however, go unchallenged by the miners. Unrest over the compulsory, fee-for-service use of company doctors led to demands for a health care plan industry-financed but union-controlled – that is the third conception. In 1946 such a scheme was initiated and, under its purview, progressive doctors undertook research on respiratory problems, leading to the view that miners suffer from a disabling killing disease which is related to exposure to coal-dust, that is an occupational disease. The picture though is more complicated than this progression from unknown through individual to economic causation suggests. To compete with oil and natural gas, the union has supported the introduction of underground mechanical loading and continuous mining technology, both of which increase dust levels and, therefore, the rates of respiratory disease. Thus, in 1968, the mining community of West Virginia started the black lung movement based not on company or union perceptions but on the collective experiences of miners and their families. While such a movement is a product of history, it used its members' experiences to challenge that history. Changing conceptions are, therefore, inextricably tied to social practice and experience, and it is through experience that groups and individuals become aware of the limitations of other conceptions and practices. Thus, for example, the efficacy of universalism is confronted by the lives of those in deprived neighbourhoods while the isolation of medical practice is challenged by healthism and that of the individual by awareness of environment and social context.

Conclusion

We have examined three broad conceptions of health and health care – the physical, the mental and the social. The isolation of key factors from these conceptions – the environment, moral adequacy, individual attributes and responsibility – provides us with important pointers for the ensuing chapters. The attempt to get behind the conceptions to discover common assumptions and characteristics will serve in a similar fashion. This attempt enabled us to discern the apparent independence and isolation of the medical sector in conventional analyses. The emphasis on the scientific treatment of the individual suggests that the malfunctioning of body and mind are amenable to technical solution. There are parallels in the provision of care. The malfunctioning of the care system is seen as treatable by technical adjustments which lend themselves well to modelling solutions (see Chapter 4). The discussions of medicine as social control and of 'health' as a commodity place the relationship between medicine, health and society at the centre of analysis. In the following chapters, we shall at times suspend consideration of these underlying forces to examine medical geography and studies of health facility location unproblematically and in their own terms. We add, therefore, a scientific, technical appraisal to the conceptual one which will reappear more explicitly in Chapter 6.

Further, so far we have looked overwhelmingly at capitalist society, and at the urbanised section of that. In Chapter 6, we will extend our analysis of these relationships to other societal contexts through a consideration of their systems of care. Indeed, it is one of the tenets of this book that the system of care prevalent in a society is in large measure shaped by the conceptions of health, illness and justice in that society. Unless that sounds too idealist, we also see history and economy as shaping the health care system, though not in any consistent ways. We thus do not subscribe to the logic of industrialism argument, that suggests that the welfare state is necessarily an outcome of industrialisation. Thus Bell (1960, 402–3) argues that 'there is, today, a rough consensus among intellectuals on political issues: the acceptance of a welfare state; the desirability of decentralised power; a system of mixed economy and of political pluralism'. Such inevitability denies the particularistic tensions between state, capital and social groupings in different countries. These points are taken up in Chapter 6. They also re-emphasise the tension between the legitimising and enabling social institutions like medicine and the experience of specific social groups. This chapter has stressed the control and integrative functions of medicine acting to maintain and legitimise systemic imperatives. Not forgetting the black lung example, we shall leave our tentative statements on lay concepts and experience as challenges until the final chapter.

MAN, DISEASE AND ENVIRONMENTAL
ASSOCIATIONS: FROM MEDICAL GEOGRAPHY
TO HEALTH INEQUALITIES

In this chapter, we shall examine the dominant concerns of geographical research in the health field, namely the relationships between medical phenomena and environment. These are also the concerns of epidemiology, which has been described as the study of health and disease in populations (Morris, 1975).

Within the epidemiological framework, Barker (1982) distinguishes between descriptive and analytic studies. The former 'are carried out in order to determine the frequency of a disease, the kind of people suffering from it, and where and when it occurs', while the latter 'are carried out to test hypotheses about influences which determine that one person is affected by a disease while another is not' (Barker, 1982, 6 and 8). Most of the studies to which we refer fall into the first of these categories, utilising cross-sectional data relating to one time period to highlight both geographical variations (in say the distribution of colonic cancer) and differences between social classes.

Medical geography can be viewed as being concerned with the spatial aspects of this branch of epidemiology. It is necessary, however, to distinguish further between medical geographical studies, with some describing disease patterns and their diffusion and others searching for causes by examining the relationships between spatial patterns and other variables. We assess the contributions made by these approaches to the study of physical and mental health, giving only brief space to the concept of social health, important as it is in associating health with quality of life. Ultimately our discussions of physical and mental health also extend beyond the geographical traditions we have identified. The search for explanation of the differential association between man, disease and environment will require consideration of health inequalities and their social and economic causation.

Physical Health

The major focus of geographical research has been on physical health, particularly the relationship between man and disease (ill-health) as mediated by the environment. 'Traditionally, this concern focuses upon the description of patterns of ill-health and mortality, on the recognition of physical and human environments which appeared to be associated with these patterns and on the ecologies of specific diseases' (Phillips, 1981, 1). In other words, the thrust of these studies was primarily ecological and epidemiological. Thus, the association between disease and environment and the initiation and diffusion of diseases form the basis of, what may still be called, medical geography. It is seen by some of its practitioners as a tool of medicine. Thus, McGlashan (1972, 14) argues that:

> Medical geography is a tool and but rarely an end in itself. It is the application of geographical methods and skills to medical problems. One may consider geographical *evidence* on medical hypotheses. It would be improper to claim that geography provides *proof*.

It is possible to illustrate such a contention from McGlashan's (1966) own work. This demonstrates that the geographical association between phenomena (e.g. male circumcision and cancer of cervix uteri in Central Africa) may be non-causative but it does suggest the bases for further work to examine this relationship.

Medical geography is, therefore, more than the geography of diseases (Learmonth, 1975).

> In brief, the medical geographer's tasks are to prepare and collate disease data and to map them to show where a certain condition is rife (or absent); to apply objective statistical tests to these distributions to assess whether or not the pattern is likely to have occurred by chance; to measure the degree of co-extensiveness between disease and other spatially varying factors; and then to apply tests to decide whether any spatial associations he has shown could be causative. (McGlashan, 1972, 12)

What, in effect, McGlashan seems to be saying is that medical geographical works sit well with the dominant paradigms of general geographical endeavour – regional geography and areal differentiation with their associated mapping techniques and locational analysis with the associated statistical techniques. There are, however, good reviews of medical geography from such a technical perspective (McGlashan, 1972; Learmonth, 1975; Pyle, 1979). We have chosen to divide our discussion into two parts and focus on the description and spatial diffusion of disease and aetiology of diseases, though it is important to point out that this distinction is more for the purposes of analysis than one which reflects a clear-cut division in reality.

Disease Patterns and Diffusions

Infectious diseases have formed the focus of studies in this category, medical geographers making a distinguished contribution to the study of the dynamics of some of the most prevalent diseases affecting human populations. As Pyle (1979, 19) observes, 'Perhaps as much as two-thirds of all human illness can be attributed to micro-organisms concerned with their own biological survival in human tissue.' These parasites colonise a wide variety of hosts and use a variety of transmission mechanisms. Transmission is either nonvectored (direct from host to host) or vectored (through an intermediary). With vectored diseases it is possible to distinguish between mechanical vectors which carry the disease externally and which are not essential to the host lifecycle and biological vectors in which the reverse is true (see Knight, 1974). Perhaps the best known vectors are flies, ticks and lice. Learmonth's (1957) study of malaria in India demonstrated that the disease is endemic, although this pattern has now been altered by the application of DDT. He also showed that several different mosquitos play important roles as vectors. Indeed, in the Punjab, Learmonth pointed out that the main vector is *Anopheles culicifacies* which breeds only in clean water. The purification of water-supplies does not, therefore, provide the whole answer to malarial control. This example demonstrates two things. First, it points to the importance of map comparisons in medical geography as the maps of malaria distribution and population change demonstrate (see Figures 3.1 and 3.2; also Stamp, 1964). Secondly, it gives additional weight to Pyle's (1979, 23) view that the geography of vectored diseases must 'develop an understanding of the complex interrelationship among agents, vectors, hosts and environmental risk factors'.

Pyle himself attempts to develop such an understanding with his case study of Rocky Mountain spotted fever. This illness is a tick-borne fever belonging to the rickettsial disease group. It is spread to

Figure 3.1: Malaria Distribution in Bengal

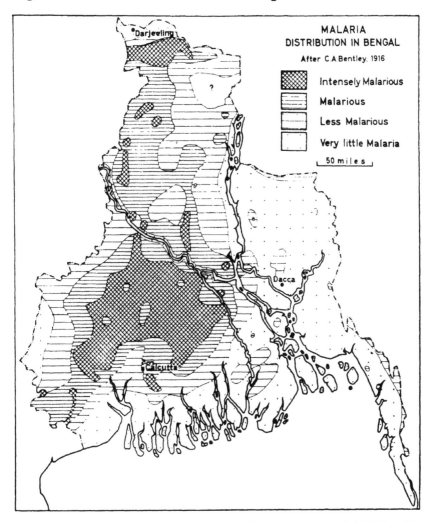

Source: Learmonth (1957), p. 51.

humans by bites from ticks which have rodents and domestic dogs as their agents, and its symptoms include high fever, skin rash and headache. Its endemic areas have shifted from the western to the eastern United States, especially the eastern piedmont, the Appalachians and parts of Oklahoma. There has also been a shift in those affected with more young and more female sufferers. In the eastern states, the disease is more prevalent in settled than wild areas and more suburban than urban. Abandoned fields and wooded areas, particularly in the piedmont, appear to be

Figure 3.2: Population Change in Bengal, 1901–11

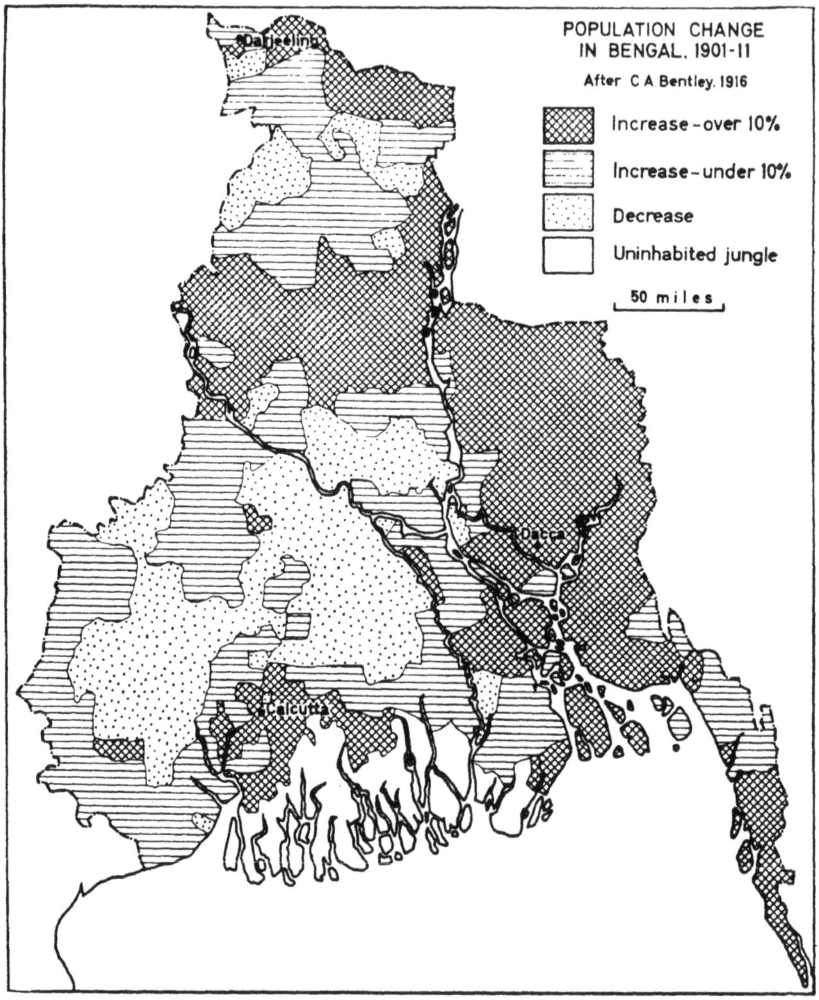
POPULATION CHANGE
IN BENGAL, 1901-11
After C A Bentley. 1916
Increase-over 10%
Increase-under 10%
Decrease
Uninhabited jungle
50 miles
Darjeeling
Dacca
Calcutta

Source: Learmonth (1957), p. 52.

geographic locations for the breeding of ticks. In Pyle's study of South Carolina, such areas are also associated with increased population growth and suburbanisation – and a recognition that domestic pets replace wild mammals as the reservoirs of Rocky Mountain spotted fever.

Such interrelationships can also be shown in another vectored disease – typhus. Typhus is highly fatal in its epidemic form and is transmitted from host to host by the body louse. These interrelationships are eloquently summarised by Howe (1972, 159):

lice took up the causative micro-organism *Rickettsia prowazekii* from the blood of people sick with the disease, and were themselves fatally infected in the process. The lice had about a week in which to transfer the infection to another subject before they died. Predisposing environmental conditions providing stimuli for the disease included the crowding together of poor, undernourished, unwashed and filthily-clad people.

It was common in prisons, and could ravish cities as in the case of London in 1741–2 where the combination of a long, hard winter, dry spring, hot summer and inadequate harvest contributed to the spread of the disease. The conditions were, however, rife for typhus in the early industrial cities of Britain. It was the poor man's disease – an unerring index of destitution – the product of squalor, insanitation, overcrowding and verminous conditions: a product of working-class housing. There were epidemics in 1817–19, 1826–7, 1831–2, 1837 and 1846–8. In 1837–8 typhus killed over 6,000 people in London in 18 months. In 1846, 1 death in 4 in Glasgow was the result of typhus alone.

Girt (1974) has argued that the transmission of nonvectored diseases – influenza, cholera, smallpox, measles, infectious hepatitis – is normally from one human to another. Pyle (1979) has synthesised much of the available information on influenza. It is primarily a respiratory ailment affecting cells in the lining of the nose, throat, trachea and lungs. The disease has regular patterns of epidemics every two to three years and of pandemics on average about every two-and-a-half decades. It is a global phenomenon made all the more virulent by antigenic variations which means that antibodies developed through past infection become powerless.

The great epidemic of 1918, for example, was responsible for 10 to 20 million deaths, many of its victims being healthy adults between the ages of 20 and 40. It has been suggested that the epidemic started in the United States and possibly spread by being taken to the European battlefronts by American troops. Using contemporary summaries of weekly cases of influenza reported in 50 major American cities, Pyle reconstructs the pattern of diffusion. He produced overall spatial patterns that approximate when different parts of the US reached the 25th, 50th and 75th percentiles of the distributions. Figure 3.3 shows the spread of influenza from east to west with suggested endemicity in Arkansas, Mississippi and Louisiana. By the tenth week the epidemic had diffused into the centre of the US with an outlier on the West coast. By the twentieth week the disease had spread to most of the country with the centre suffering prolonged effects.

At a more localised level, further work by Pyle (1973) can be used to illustrate the diffusion and patterning of measles in Akron, Ohio. Pyle calculated the age-specific attack rates for children aged under ten years for 1965, 1966, 1969 and 1970. It was found that the highest attack rates were concentrated in the central and south-eastern parts of Akron (the poverty areas). But the number of potentially vulnerable children was found to be highest in the higher income suburban fringes. This finding was explained by analysis of the 1970 epidemic. Pyle found that the epidemic originated in a poverty area in south-east Akron (Figure 3.4). Thereafter clear patterns of contagious diffusion were observed, first in the city core, and then in the transitional residential areas and finally in suburban districts. During the last three months of the epidemic the number of cases lessened because of natural causes and inoculations. Many of the cases in the final generation of the outbreak, however, were pre-school children in the initial generating area, reflecting patterns of measles diffusion which preceded the availability of vaccine. Further, the vaccines were relatively expensive. Price was therefore a barrier to the diffusion of inoculations to the entire population at risk. We can, therefore, see that Pyle's analysis has necessarily gone beyond the patterning of measles and has started to look at the disease's aetiology (its causation), not only in a medical sense but also in social and cultural ways to explain the differential diffusion of diseases among populations.

We can further illustrate the diffusion of nonvectored, infectious diseases by looking at cholera. Man is the prime reservoir for cholera. The cholera vibrios which spread from one individual to another are carried initially by human faeces which contaminate water. They are passed from man to man 'either by stools which contaminate clothing, linen, or the hands, allowing transmission through contact, or more usually through the intake of water or food contaminated by the excrement of cholera patients' (Howe, 1972, 179). Cholera is extremely contagious and in the eighteenth century was diffused through Burma, China and Sri Lanka by pilgrims, traders and military personnel. Its endemic source lay in Asia and during the nineteenth century several pandemics originated from Bengal and Indonesia and secondarily from India, Pakistan, Burma and Southern China. Cholera reached world

Figure 3.3: The Diffusion of Influenza in the United States in 1918

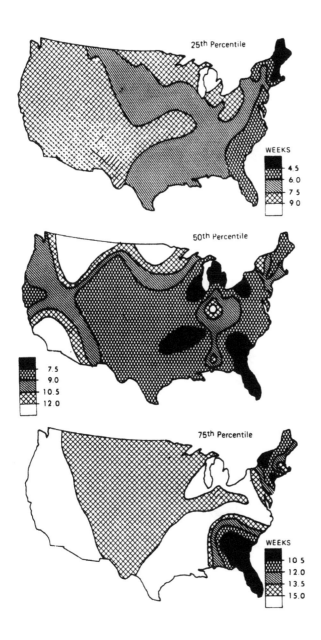

Source: Pyle (1979), p. 50.

Figure 3.4: Measles Diffusion in Akron, Ohio

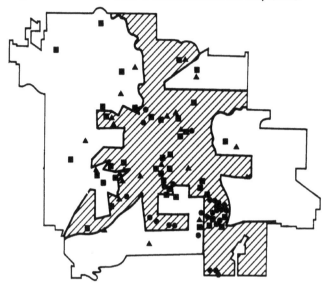

- ● 1st generation, 12/9/70
- ■ 2nd generation, 12/21/70–12/31/70
- ▲ 3rd generation, 1/11/71–1/21/71

▨ Akron's poverty areas

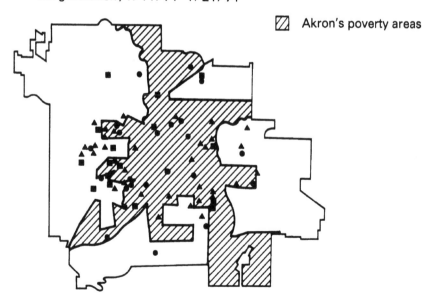

- ● 3rd generation, 1/11/71–1/21/71
- ■ 4th generation, 1/22/71
- ▲ 5th generation, 1/25/71

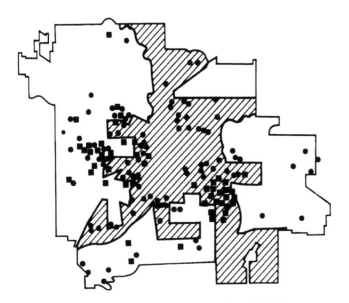

- 7th to 10th generations, 2/11/71–4/5/71
- 11th to 14th generations, 4/7/71–5/28/71

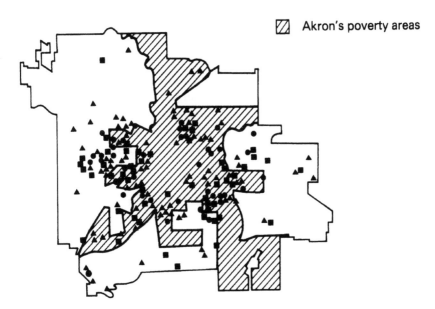

Akron's poverty areas

- 5th generation, 1/25/71
- 6th generation, 1/28/71–2/10/71
- 7th to 10th generations, 2/11/71–4/5/71

Source: Adapted from Pyle (1973).

significance with the pandemic starting in 1817. It originated in India, moved east to China and the Phillippines, south to Mauritius and Reunion and north-west to Persia and Turkey. By 1826 a pandemic covered the whole of China, Japan and Asiatic Russia (see Cartwright, 1977). In 1829 Poland, Germany, Austria and Sweden were infected; by 1831 Great Britain; by 1832 Quebec, New York and thence to Mexico.

The rapidity of the diffusion in the British Isles has been documented by Howe (1972). It appeared in Sunderland in October 1831 and thence spread to Tyneside, moving north through Northumberland reaching Edinburgh in January, Glasgow in February and Inverness in May 1832. Another wave moved south to Leeds, York, Liverpool and Manchester–Salford. London was reached by February, the Midlands in June and the West country in July (see Figure 3.5). The conditions of the industrial era were fertile grounds for the spread of cholera. The crowding and insanitary conditions of the slums helped the spread of the disease. Howe (1972, 181) cites the contemporary report of Shapter of Exeter:

> This inadequate water supply combined with the deficiency of drainage, is of itself sufficient evidence, that the necessary accommodation for the daily usages of the population must have been very limited . . . they speak of dwellings occupied by from five to fifteen families huddled together in dirty rooms with every offensive accompaniment; slaughter-houses in the Butcher Row, with their putrid heaps of offal; of pigs in large numbers kept throughout the city . . . poultry kept in confined cellars and outhouses; of dung-heaps everywhere.

Cartwright (1977, 95) re-emphasises the point: 'Few families occupied more than a single room . . . Cleanliness, privacy, decency, proper sanitation and water-supply were all impossible . . .' Privies remained uncleared, closes overflowed, the stench of bad drainage was everywhere, even among the urban wealthy in, for example, Belgravia. Thus, Howe and Cartwright are beginning to point to the causes of the spread of the disease. Indeed, as we previously averred, it is impossible to separate pattern from cause, diffusion from explanatory mechanism.

Figure 3.5: Progress of the 1831–2 Cholera Epidemic Through the British Isles

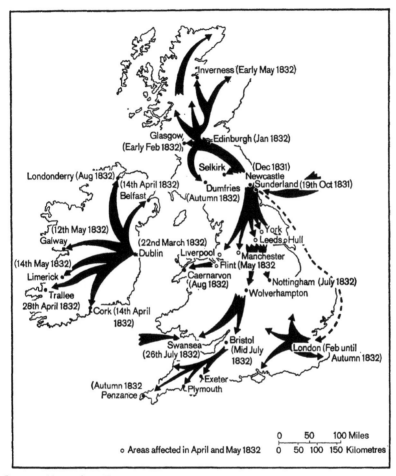

Source: Howe (1972).

Aetiology: The Search For Causes

We have seen in our look at some 'epidemiological' studies that examine the relationship of man and environment through the patterning and spread of disease that this approach does not stop at the level of description. Indeed, many such studies suggest associative relationships, point to mechanisms that make certain populations more disease-prone than others and begin to speak of causes. A separate branch of medicine, however, looks at the causes of diseases, that is aetiology. At one level, this is a medical concern – a search for the pathogens that cause disease and disease diffusion. At another, it is an attempt to explore the links between disease and environmental correlates, an approach we have already met, which is called 'associative occurrence' by McGlashan (1972). At a further level, it looks at the differential rates of occurrence among social groups and territories to suggest that social, economic and cultural factors intervene in the associative relationships between disease and environment. This third level is to anticipate for it takes us from the realm of medical geography and into that of health status and societal constraints (see below).

The search for causes is closely linked to the disease ecology approach of May (1950; 1961), which is primarily concerned with the relationship of man with his total environment. Pyle (1979) points to two major elements in May's work – the understanding of disease *hazards* in relation to the natural environment and the ecology of vector-borne diseases. Indeed, with the latter element, we have already seen how this enables us to link epidemiological and aetiological concerns. In essence, the disease ecological approach attempts to sort out and fully understand distinctions between biological and cultural determinants of human diseases and – as Voronov (1977) has suggested – isolate the dominant causes of disease in relation to landscape, by seeing environment as an indirect cause of disease.

Thus, May saw the relationship between environmental factors (geogens) and the pathological, causal factors (pathogens) as crucial. May identified five pathogens: the causative agents (viruses, protozoa, rickettsiae); vectors (which may spread the causative agents to man – flies, ticks, lice); intermediate hosts (organisms essential to the lifecycle of the agent); reservoirs (hosts which carry the infection in nature until it is picked up by man); and finally, man himself (often the most important reservoir). But the impact of these pathogens depends on those kinds of environmental factors – the geogens. These are:

1. Inorganic stimuli which include physical environmental features such as heat, soil, water and wind. May considered climate to be the most important of these inorganic stimuli, directly through the debilitating effects of excessive exposure to high temperatures and humidity or indirectly through particular levels of temperature and humidity favouring the development of particular parasites. Mills (1944), for

example, points to the close relationship between stormy weather and respiratory infections. Stamp (1964) was, however, correct in pointing out that we spend a great deal of time indoors in artificially controlled climates and that the impact of climate on health may, therefore, be reduced (at least for certain groups).

2. Organic stimuli – closely linked to inorganic ones including animal and vegetable life and parasites. Particular plant and animal ecologies at specific altitudes in specific microclimates can lead to a proliferation of different disease-types. Pyle (1979) exemplifies the importance of ecological niches with reference to Burkitt's lymphoma – a jaw tumour – with a high prevalence in tropical Africa and New Guinea. In Africa a lymphoma belt can be said to straddle the Equator. It appears that rainfall and altitude can help explain the presence or absence of the disease, in East Africa there being an altitude barrier at 3,000 or 5,000 feet depending on which authority is accepted (see McGlashan, 1969). Further, Roundy (1976) has pointed to the limiting effect of altitude on the endemicity of such diseases as schistosomasis and malaria in Ethiopia. He does suggest that human movement for farming, grazing and social gatherings may alter the distribution of the diseases.

3. Sociocultural stimuli – population distribution, diet, culture, housing, sanitation, customs, beliefs – can also cause disease or at least alter its distribution. May's own work did not stress these factors to any great extent and Phillips (1981) is right when he suggests that there are strong elements of environmental determinism in his work. Man's diseases were seen as a direct result of the physical environment within which he lives, although May did see man as being pathologically tied to his culture. In fact, it is possible to note strong parallels between the work of May and that of Park, Burgess and the Chicago school which has been so influential in the development of urban and social geography.

While Pyle (1979, 83) is correct in suggesting that studies on international health problems have failed to pay sufficient attention to environmental complexes in the causation of disease, work on environmental association has a long and distinguished history. Indeed, one of the earliest and still most famous examples of such study is that of John Snow in the 1850s. Snow was a general practitioner in the Golden Square area of Soho, London. Within ten days in late August and early 1854 over 500 people died of cholera. Snow plotted the distribution of cholera deaths and discovered that the vast majority lived in an area bounded by Great Marlborough Street, King Street, Brewer Street and Dean Street and focused on the water pump in Broad Street (see Figure 3.6). Snow showed that most of the deaths occurred among those who drank water from this pump whereas those living in the same neighbourhood and using different pumps escaped. Examination revealed that the water at Broad Street had become infected by seepage from a leaking cesspool or drain. Snow had the handle of the pump removed, although this was largely a symbolic gesture as the outbreak was already limiting itself. He did demonstrate the association between cholera and contaminated water.

A more recent demonstration of environmental association is provided by Howe (1960) and his work on cancer in Wales. He demonstrates that large variations exist in the incidence of mortality from different cancers. His conclusion is rightly tentative:

> The pattern of distribution of mortality from lung-bronchus cancer in males is the only one which offers sufficient correlation with the pattern of selected indices of atmospheric pollution to suggest an association with atmospheric carcinogens . . . in the causation of stomach cancer the untreated acidic water-supplies in daily use in the countryside, and in particular those waters polluted by effluent from spoil mounds of defunct lead, zinc, and copper mines . . . might be worthy of further study.

Sorokina (1976) found in the Moscow Oblast that lung cancer mortality was virtually identical for both urban and rural areas. For the rural areas, she found that many of the factors positively correlated with such mortality – viz. use of farm pesticides from the carcinogenic dithrocarbonate group, percentage of workers employed in agriculture, the amount of smoking *per capita*, the amount of dust-causing ploughing, the use of farm machinery (exhaust fumes) and the use of kerosene (producing household soot) – were associated with the environment.

The above examples are merely illustrative of that branch of medical geography that has been concerned with the search for associative occurrences (see McGlashan, 1972; Pyle, 1976),

Figure 3.6: Deaths from Cholera in the Soho District of London, September 1854

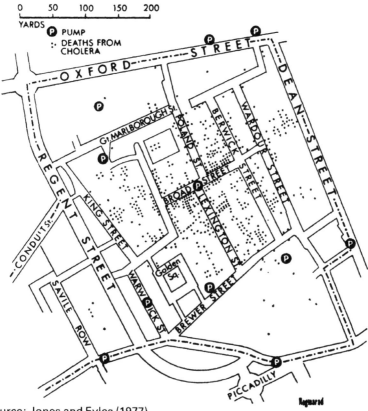

Source: Jones and Eyles (1977).

whereby an attempt is made to associate disease and certain environmental factors through statistical analysis utilising an array of techniques from the relatively uncomplicated to the more sophisticated. Thus, Girt (1972) examined the incidence of simple chronic bronchitis in Leeds. He argued that as the disease is the result of interactions between the individual and his environment, the risk of developing it will increase with lengthening exposure to a causal factor in the environment. He obtained a representative range of environments by dividing Leeds into 30 quadrats based on the ecological models of Burgess and Hoyt (i.e. concentric and sectoral zones). He interviewed 20 randomly selected women in each quadrat. He found that there was a strong positive correlation between bronchitis and overcrowding and dampness in housing. On this basis the disease probably shows a strong sectoral component, highlighting working-class districts in the inner and transitional areas and individuals who have moved but who have a long history in such housing. This pattern is though greatly complicated by cigarette smoking which significantly alters the relationship between bronchitis and ecological structure. Dever (1972a, b) points to a tentative positive relationship between leukaemia and poor, overcrowded housing in Buffalo and Atlanta while, in a more recent study, Giggs *et al.* (1980) using probability maps and map comparison suggest that there is an association between the incidence of acute pancreatitis and the chemical composition of water supplies in the Nottingham area.

A more sophisticated associative study employed factor analysis (Pyle and Rees, 1971). This study used 18 disease variables along with population density for 76 community areas in Chicago. The first factor (39.8 per cent of the variance), when mapped, produced an ecological patterning of socioeconomic status (see Figure 3.7). It was labelled the poverty syndrome and demonstrated a close correspondence between the incidence of tuberculosis, gonorrhea, syphilis and infant mortality and the poorest neighbourhoods in the densely populated inner city. The second factor (22.7 per cent) – a density syndrome – had associations with three childhood diseases: mumps, whooping cough and chickenpox, and was linked again with poor neighbourhoods. The third (14.0 per cent) and fourth (13.1 per cent) factors: the upper respiratory and rubella syndromes – showed no clear geographical pattern; but the fifth (10.4 per cent), a water syndrome, isolated infectious hepatitis and identified a definite relationship between high factor scores and stretches of open water. It must be stated that even where methodologically sound, such studies can seldom isolate causative mechanisms. In an attempt to improve explanation, Pyle and Lauer (1975) and Pyle (1979) have used canonical analysis and associated stepwise regression. Even then, as Pyle admits, little has as yet been added to clinical knowledge, with his study of Ohio demonstrating the increasing incidence of heart disease with increasing age and lower income.

With the search for causes, we have come full circle in our examination of the relationship between man, disease and environment. The search for pattern is a necessary component of any search for causation. To know the initiation and diffusion of disease is a vital component in understanding its causation and assisting in its amelioration. Indeed, Morris (1975, 142) suggests

Figure 3.7: Disease Mortality and Morbidity Syndromes in Chicago

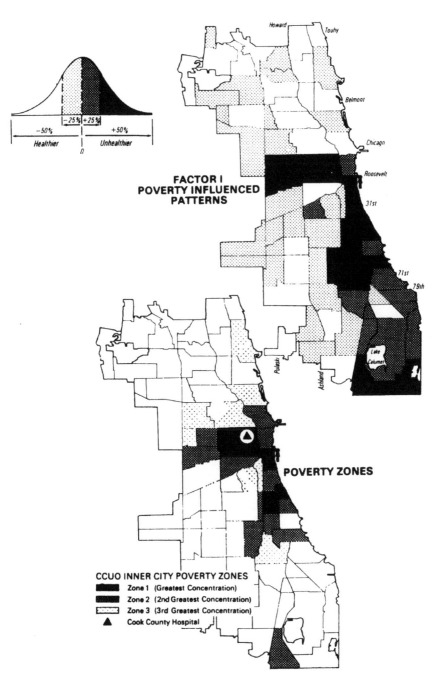

FACTOR I
POVERTY INFLUENCED
PATTERNS

POVERTY ZONES

CCUO INNER CITY POVERTY ZONES
- Zone 1 (Greatest Concentration)
- Zone 2 (2nd Greatest Concentration)
- Zone 3 (3rd Greatest Concentration)
- ▲ Cook County Hospital

Figure 3.7 (continued)

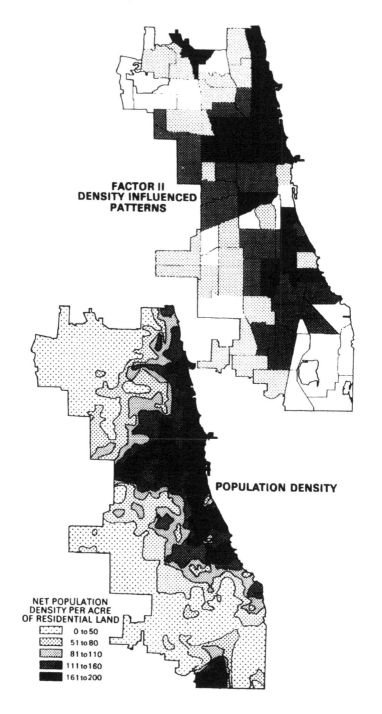

FACTOR II
DENSITY INFLUENCED
PATTERNS

POPULATION DENSITY

NET POPULATION
DENSITY PER ACRE
OF RESIDENTIAL LAND

0 to 50
51 to 80
81 to 110
111 to 160
161 to 200

Figure 3.7 (continued)

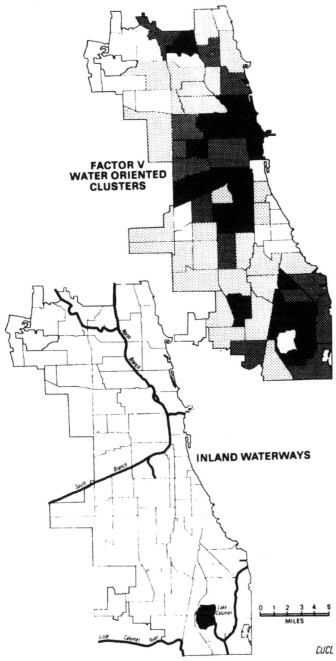

FACTOR V
WATER ORIENTED
CLUSTERS

INLAND WATERWAYS

0 1 2 3 4 5
MILES

CUCL

Source: Pyle and Rees (1971), p. 351.

that 'the main use of epidemiology is to discover causes of health and disease so as to increase understanding, and hopefully also help improve the conditions of the people'. Soviet medical geography has had as its main aim, in fact, the elimination of diseases and as such has been part of health care policy. Thus, according to Shoskin (1964), not only has Soviet medical geography evaluated natural conditions and landscape regions in relation to disease and studied the endemicity and diffusion of diseases, but has also utilised its evaluations to effect public health programmes. Environmental and climatic information were necessary elements to planning the Ust'-llimsk industrial area (Soboleva, 1974). Thus, prior to the influx of industry and population after the construction of a hydroelectric power station, the probable effects of the harsh climate on humans were explored while the geochemical environment, in particular the known shortages of iodine and fluoride, was also examined.

Morris (1975), however, is surely correct in pointing out that the association of disease and environment is only the beginning of the discovery of causes, which involves comparing the health and diseases of groups defined by composition, inheritance, experience, behaviour and environment. Such a search takes us beyond the realm of traditional medical geography but it does place geographical distribution and environmental factors into a broader framework. Such a framework may well suggest not the unimportance but the continued relevance of a modified medical geographical approach. We can examine this contention with reference to cancer mortality in Great Britain.

Gardner *et al.* (1982) suggest that the use of large areas – counties, regions – in studies of cancer mortality may mask the presence of aetiological factors in the local environment. In other words, the detection of such factors is in part a matter of scale of analysis. Using data for local authority areas, they examine the variations in cancer mortality. They discovered that the mortality ratios from pleural mesothelioma were particularly high in eastern London, centred on Barking, around Merseyside, Tyneside, Southampton, Barrow and Plymouth. This association with the main dockyards and ports points to the extremely high use of asbestos, particularly crocidelite, in shipbuilding and repairing in the relevant past. Gardner *et al.* also associated high rates of cancer of the nose, nasal cavities, middle ear and recessory sinuses in men with particular industrial environments. Thus, for example, the

High Wycombe and Tower Hamlets areas are, or were, associated with the furniture industry while Rushden (Northamptonshire) is a centre for the manufacture of boots and shoes.

As a third example, Gardner *et al.* looked at bladder cancer in men and found higher incidences of mortality in northern England (South Lancashire and West Yorkshire) and in and around London. The dyestuffs and rubber industries are concentrated in many of these areas. We can take the consideration of bladder cancer somewhat further by referring to the 'detective' chain described by Morris (1975). In 1895, bladder cancer was first described among a small group of workers engaged in the manufacture of magenta from aniline, which was incorrectly diagnosed as the most likely agent. During the 1914–18 war, the manufacture of dyestuffs increased in Britain as German supplies were halted. The health problem was not, however, revealed by national occupational mortality statistics because the population-at-risk was small and included in other groups. (Classifications may hinder the search for associations and causations.) Tumours did develop though at younger ages than for non-occupational cancers, having an incubation period of 15 to 20 years and an individual risk of one in five to ten. Indeed, for one small group distilling β-naphthylamine the risk was one in one. In fact, in 1948 this product along with x-naphthylamine and benzidine were found to be responsible for producing the cancer. But the cancer was also found in rubber goods workers who were used as a 'control group' and an analysis of death certificates found cable workers had an excess of the disease, rubber being used as insulation on cables. While the manufacture of β-naphthylamine was halted in Britain in 1952 and screening introduced, cases of bladder cancer still occur in rubber and cable workers. The causes are still not fully known.

This example provides a piece of epidemiological and aetiological detective work. It suggests that the detailed geographical analysis of mortality may detect associations between cancers and known carcinogens. It also points to the close association between disease and work environment and to the importance of linking statistical analysis with clinic observation to discover the aetiology of disease. It may be that such an example forms a model to be followed in the search for associative occurrence, although it takes us well beyond the geographical realm to consider differential risk and the social and economic contexts of ill-health (see below).

Mental Health

The study of mental illness does not lend itself as readily as that of physical disease to a treatment based on the examination of man–environment relationships. Where such study has been carried out, an ecological perspective has usually been adopted. Elements of the social environment are thus selected and averaged for particular spatial units and then related to rates of specific mental disorders. Such analyses are, therefore, associative, similar to those discussed in the previous section.

Dunham (1937), for example, analysed the distribution of schizophrenia and manic depression in Chicago in the 1930s. His analysis of schizophrenics involved 7,253 cases in 120 community areas in which incidence rates varied from 111 per 100,000 on the urban fringe to 1,195 per 100,000 near the central city. The community areas with the highest incidence rates were all found in the central city and corresponded with hobohemia, the rooming-house district and the ethnic quarters. A regular incidence gradient, with scores decreasing from centre to periphery, was found to exist even when race was held constant. In his analysis of manic depression, Dunham failed to identify a spatial distribution which could be described as other than random. Manic depression appeared to be more related to personality and psychological factors than to the urban environment. The strong relationship that schizophrenia bore to ecological structure was, however, replicated by the incidence of senile psychoses, alcoholic psychoses and drug addiction (see Figure 3.8 and Faris and Dunham, 1939). All these disorders were concentrated in the poor, dilapidated parts of Chicago. To be more specific, paranoid schizophrenia was associated with the room-house districts, catatonic schizophrenia with neighbourhoods of first immigrant settlement, manic-depressive psychoses with areas of high rents, alcoholic psychoses with rooming house and certain immigrant districts, dementia paralytica with black communities and senile psychoses and arterioschlerosis in districts with the lowest percentage of home-owners. Support for the association between schizophrenia and ecological structure was found in a more recent study by Mintz and Schwarz (1964). They also confirmed the spatial randomness of manic depression.

The inner city location of high incidences of schizophrenia as well as other mental disorders has been found in Britain, as studies

Figure 3.8: The Distribution of Severe Mental Disorders in Chicago

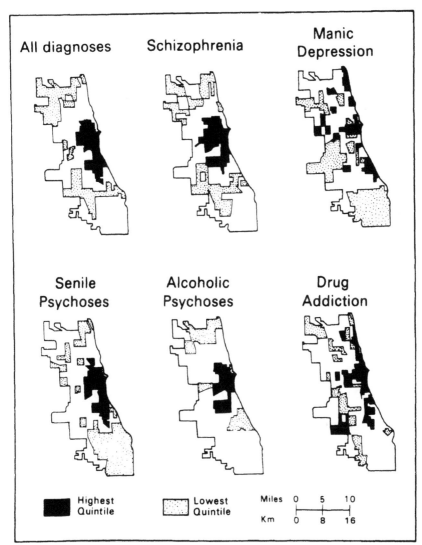

Source: Giggs (1979).

of Liverpool (Castle and Gittus, 1957) and Luton (Timms, 1965) indicate. Giggs' (1973) study of schizophrenia in Nottingham also substantiated the established pattern. He collected data on the 444 individuals classified as schizophrenic and admitted to hospital from Nottingham addresses for the first time during 1963 to 1969. A gradient analysis, based on 1 km zones, demonstrated that most schizophrenics came from in, or near, the city centre with 68 per cent of all patients coming from within 4 km of the centre. He concluded that:

> the rates of schizophrenia are closely correlated with those for a whole set of unfavourable life circumstances, notably low social status, high unemployment, and low social cohesion (indexed by high rates of spatial mobility and social isolation, minority group status, family disruption, and single-parent households). (Giggs, 1973, 71)

Such circumstances are seen to assume their greatest intensity in the inner slum districts of cities. This view of a central location for mental disorder – broadly conceived to include stress and distress – is given direct support by Srole (1962), who considered that 81.5 per cent of the sample of people in Manhattan were mentally diseased to some degree, the vast majority in the lower socioeconomic groups, and by Bagley *et al.*'s (1973) study of Brighton; and indirect support by Bastide (1973), who argued that mental disorders are higher for single persons and those who have experienced traumatic events than for married people, there being an overrepresentation of the former groups in the inner city. We shall not repeat our discussion of the problems of defining mental illness (see Chapter 2) or make much of the fact that the questions asked concerning stress are likely to determine which social group is going to appear to be most 'mentally diseased' (see Hartung, 1963). We must, though, look at the studies which challenge the association between central city and mental disorder.

Even the early studies (see Faris and Dunham, 1939) noted substantial changes in the distribution of several mental disorders over a period of 40 years. These shifts have been recently substantiated for Chicago by Levy and Rowitz (1973). Further, both Castle and Gittus (1957) and Timms (1965) noted the association of mental disorder with peripheral public housing estates as well as with the central city. While it is important to remember that the cases of mental disorder in any neighbourhood are far outnumbered by those who classify themselves (and are classified) as mentally normal, it may be that the social milieu of particular neighbourhoods rather than ecological structure *per se* results in a higher than average incidence of mental disorder. Indeed, the movement of population from inner city to peripheral housing estates may have the effect of redistributing patterns of mental illness. If that is the case, the explanation of mental disorder is to be found in the personality structure of individuals with predisposing potentialities being present in the social structure of certain communities.

Support for such explanations can be found in the drift hypothesis. Giggs (1973), on the basis of his findings on schizophrenia, suggests that there may be pathogenic areas in cities which seem to destroy mental health. Indeed some environmental settings *may* create schizophrenia while physical changes to the urban fabric – renewal – may alleviate it. Such a view not only suggests an environmental determination of schizophrenia but also that social and mental improvement may come through physical reconstruction, an idea prevalent in the nineteenth century when bad behaviour was associated with the poor sanitation. Further, in a critical review of Giggs' views, Gudgin (1975, 148–9) suggests that he:

> appears to ignore an intuitively more plausible reason why schizophrenia should be much more prevalent in the central areas of cities. It is possible that sufferers move into areas of poor quality housing and may thus not be a product of that environment in any meaningful sense.

In the inner areas such people will inflate local hospital admission rates and rates of mental disorder and social deprivation. They drift into the inner city which does not produce them. They simply collect there, psychologically vulnerable people drifting in from random locations. The environmental association becomes more tenuous, the relationship mediated by the notion of personal disorganisation. To avoid a psychologistic explanation it is possible to relate mental disorder to culture and social structure. It is possible to speculate that the vulnerable drift into the inner city because they are unable to compete in other housing markets for living space and to cope with a social order that values material success and acquisitiveness highly. Further support for a cultural as well as psychological explanation of mental disorder can be found in the study of suicide.

Suicide is not usually regarded as a mental disorder or disease of the order of schizophrenia and manic depression. It is, however, often the sad finale for such sufferers. Further, it may be regarded as the absolute abnormality. If those defined as mentally ill are in part the individuals who think and structure their lives differently from some societal 'norm', then suicides take this to its ultimate conclusion; theirs is the total withdrawal from the rules governing social life – the end of their own existence. This social definition of suicide owes much to the work of Durkheim who elaborated the relationship between individuals and society (see Durkheim, 1964). How can a collectivity of individuals make up a society? Durkheim's answer was social solidarity, a view that provides a social explanation of individual attachment. In industrial society, this social explanation concerns organic solidarity in which consensus is obtained by the interdependence of different parts. As individuals, we depend on society because we depend on the parts that compose it (see Lukes, 1973). Durkheim saw such interdependence and individuality as a normal, happy development in human societies. 'He approves the differentiation of jobs, the variability and differentiation of individuals, the decline in the authority of tradition, the expanding domain of reason, the allowance for individual initiative' (Aron, 1970, 34). The individual is, however, not necessarily any more satisfied with his lot in such societies. Durkheim was in fact struck by the increase in the number of suicides and saw in this supremely individual moment the presence of society in the individual's consciousness. The rate of suicide in fact varies with the degree of integration of individuals into the group. Thus egoistic suicide is inversely proportional to the degree of normative integration to be found in the groups of which the individual is a part. There are more suicides in urban than rural areas, among the single, widowed and divorced compared with the married and among Protestants as opposed to Catholics. Opposite to this is altruistic suicide where on certain occasions suicide is demanded by the group and high social cohesion ensures that it is carried out, for example widows in India, soldiers in Japan. Finally, anomic suicide is the consequence of the weakening of social ties in contexts where there is a lack of normative definition and direction. Anomic suicide is closely related to economic crisis.

Durkheim's social morphology – the study of the *substract social* (the distribution of social forms) – was entirely social; it was the group framework of social life (see Buttimer, 1969; Jones and Eyles, 1977). Despite this, his work does form the basis of studies that examine the ecological association of suicide. Indeed, it is possible to see the antecedents of such work in the extensions of

Durkheim's analyses. Explicitly, Sorre (1957) argued that Durkheim's definition of environment was too narrow, there being many occasions on which physical conditions influence social differentiation. While such views were mainly taken up in the general field of human geography, it is possible to link them to the alternative to the drift hypothesis of mental disorder – the breeder hypothesis, alluded to by Giggs (1979), and which attributes causal roles to the poor social climate and housing conditions prevalent in inner residential areas. Thus, in Chicago, Maris (1969) found that high rates of suicide had persisted in the Loop and the central city areas of the North and West side for 30 years or more. A declining rate, however, was found in the Near South Side, an area of predominantly black population. It may be possible to explain their lower rate to their stronger group ties and attachments. In London, Sainsbury (1955) found a persistence of high suicide incidence in five central city districts north of the Thames, namely City, Holborn, Westminster, Paddington and Hampstead. The incidence declined in St Marylebone while becoming more prominent in Chelsea. In fact, both Maris and Sainsbury view suicide as a more middle-class phenomenon, with high incidence areas being typically populated by slightly more educated and higher income residents than found in other areas. Economic upheaval – unemployment, status change – was related to suicide as was isolation, suggesting that in the modern city, anomic suicide is the more prevalent type. Further, if suicide is mainly a middle-class phenomenon, but one that is more prevalent in lodging house districts with transient, isolated inhabitants, support is given to the drift hypothesis. Further support comes from its concentration in the central area of Sydney (see Figure 3.9), where death rates from other emotionally related problems – cirrhosis of the liver and peptic ulcer – are also high (see Burnley, 1977; Gibson and Johansen, 1979). As Gibson and Johansen (1979, 74) comment:

> The concentration of high suicide rates in the central parts of Sydney parallels the experience of many large cities. However, the low suicide rates in outer suburban areas are remarkable. These areas include Mt. Druitt, Auburn, Canterbury and Villawood, all of which figure prominently in the maps of depression, sleeplessness and social problems. A possible explanation is that emotional problems respond most readily to the love and care of families and close friends. In the centre of

the city single or divorced people, new migrants from outer areas and the socially underprivileged of all groups have little psychosocial support to help them cope with emotional problems.

A recent study by Bagley and Jacobson (1976) suggests, however, that different types of suicide have different ecological associations, with 'sociopathic' suicide occurring predominantly in central areas and 'physical illness' suicide in middle-class districts. We should add that this study neither substantiates or rejects the drift hypothesis, but gives further weight to our idea of psychological and cultural explanation overlying these ecological associations, which are dependent, to a large extent, on the accuracy of local suicide rates. These rates are themselves based on relatively low incidence rates. Area-based data on suicide (and on illness in general) are thus problematic: not only in the sense of the drift hypothesis, but also in the danger of seeing problems manifested in localities as problems of areas as such.

An implicit extension of Durkheim's social morphology has also had an impact on ecological studies of mental illness and stress. It is possible to argue that Durkheim's ideas on organic solidarity and interdependence, which he suggested would occur in the context of mass markets and urban growth (i.e. centralising tendencies), find expression in the ideas of Wirth (1939) on urban life. But whereas Durkheim viewed the individuality and interdependence of modern society favourably, Wirth's scheme has negative connotations with impersonality, superficiality, anonymity and conflict dominating interpersonal relationships. We saw in the last chapter how mental health as opposed to mental disorder became associated, in social thought, with particular ways of life. Perhaps it would be more accurate to say that the city became identified with ill-health. It is not our purpose to repeat these arguments, instead we want to examine those studies that have suggested that stress and depression are connected with particular urban environments. The argument is not straightforward. There is much conflicting evidence with different environments being regarded as stressful by different researchers and the same environmental conditions being viewed differently by different researchers.

It is popularly supposed, for example, that overcrowded conditions result in stressful behaviour. Indeed, Mercer (1975) points to academic authorities who regard the analogy with animal

Figure 3.9: Suicide Rates in the Greater Sydney Area

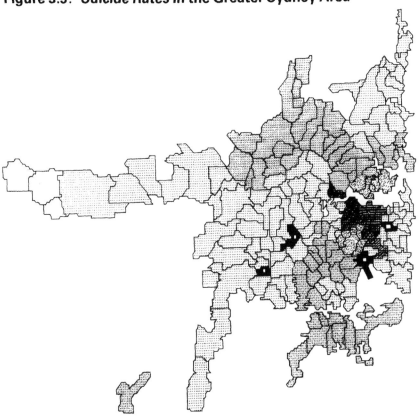

Range
- 0.05 — 0.16 per 1,000
- 0.16 — 0.26 per 1,000
- 0.26 — 0.37 per 1,000
- 0.37 — 0.47 per 1,000
- 0.47 — 0.58 per 1,000

Source: Gibson and Johansen (1979), p. 75.

populations as sound and who see apparently invariable consequences stemming from severe overcrowding. But as Fawcett
(1970) observes:

> The deleterious effects on mental health of unwanted
> pregnancies, oversized families and urban crowding have been
> commented upon by many writers. Typically, reports dealing
> with the effects of population variables on mental health are
> based upon clinical impressions or unsystematic observations
> . . . causal effects are not established and intervening
> psychological processes not specified.

Indeed, Hare (1956) in a study of Bristol shows that there is no
sufficient statistical relationship between population density and
mental health. Overcrowding *per se* seems to have no deleterious
effect on health (see Booth, 1976). Further, Biderman *et al.* (1963)
argue that 'physical crowding has significance only in
interdependent relationship with many other variable features of
the entire situation, including environmental, structural, temporal,
psychological and social features'. Schmitt's (1963) study of Hong
Kong, where some individual neighbourhoods had over 2,800
people per acre, failed to find the anticipated levels of mental illness
and social pathology. Schmitt explained this in terms of Chinese
traditions and family cohesiveness – a factor interestingly related to
Durkheim's views – low space expectations, luck, medical science
and efficient administration and a low car-to-person ratio. In any
event, social structure mediates between density and mental illness
– the ecological association is simply not there.

A variant of the population density theme is seen in studies of
housing layout and design. At various times, new towns, suburban
estates, public housing, tower blocks and the inner city have been
popularly regarded as being stressful. Taylor and Chave (1964),
however, found an equal incidence of the 'subclinical neurosis
syndrome' in a decaying London borough, an out-country estate
without local work or social life and a planned new town. They
found that those who complained of loneliness and boredom in the
new town did so *because* of their poor mental health and not vice
versa. In fact, it was not the newly arrived nor the isolated young
mother with children who suffered most. Taylor and Chave
demonstrate that the highest incidence of the neurosis is found
among women aged 45 to 54, and that it takes at least five years for

any effects of the environment to become apparent. They further suggest that the incidence of the syndrome is likely to increase if the complaint is expected and regarded as respectable. The low figures for the new town do not, however, support this conclusion. The general findings of Taylor and Chave were substantiated in a study of the old central area and a new estate of Croydon by Hare and Shaw (1965) – there was equivalent incidence.

In more recent years, much work has been carried out on the mental health of residents in high flats. Hirt (1966) found among young children (below 10 years of age) flat dwellers had twice the incidence of respiratory infections compared with children in houses. Among adult flat dwellers there were twice as many clinical consultations in which symptoms of emotional disturbance predominated than among house-dwellers. Fanning (1967) found that flat dwellers had higher attendance rates at doctors' surgeries and higher referral rates to specialists than house-dwellers. It was not, however, simply a matter of environment. Neurotic symptoms were three times as prevalent among mothers aged 20 to 29 living in flats as among the same age group in houses, whereas wives over 30 showed the same rate of incidence of psychoneurotic disorders irrespective of dwelling type. Stage in the lifecycle seems a potentially important determinant of neurotic risk. The evidence is contradictory and inconclusive, Richman (1974) finding maternal depression common in high-rise flats but Moore (1974) seeing no relationship between psychiatric illness and building height and design.

Further, Darke and Darke (1970) point to the isolation prevalent in high flats in Greater London. They comment that isolation depends on the individual's desire for privacy, although as Jephcott's (1971) study of Glasgow indicates, living alone in such accommodation can accentuate such difficulties. Social ties (and relative mobility) can, therefore, militate against feelings of isolation and depression. Pfeil (1968) in a study of Dortmund–Nordstadt found that strong neighbouring relations could result in satisfaction with high-rise living. He did point out, however, that primary relationships developed most quickly in smaller blocks where there was little movement of tenants. It may be that many of the individuals and families relocated from slum houses to high flats experience grief, stress and depression. Martin *et al.* (1957), Gans (1962) and Fried (1963) all point to the health costs of the destruction and renewal of established residential areas. Such findings may suggest that any dislocaton of an established way of life, especially if that dislocation is unexpected, unplanned or forced (i.e. beyond the individual's control) may produce stressful reactions. This concerns not only change of residence but also bereavement, unemployment and so on. We again see environmental association diminished. If stress is part of the human condition then forced change of residence is a minor contributing factor, affecting comparatively few people for a comparatively short period of time. There are, therefore, a great variety of factors explaining the incidence of mental distress, as Darke and Darke (1970, 29) suggest in relation to high flats:

> age stage in the life cycle, socio-economic status, the homogeneity or otherwise of neighbours, previous housing experience, proximity to kin and friends, personality factors, the saliency of housing in the individual's and the group's value systems, satisfaction with other aspects of living, size and age of blocks of flats, layout of the block, floor on which the dwelling is situated, detailed design of the flats and so on.

Once we admit that social and cultural factors influence the likelihood of mental distress, then the ecological pattern and environmental association are in large part functions of these societal factors. This is not to deny that the poor, the old and the ethnical minorities, may be concentrated in particular areas and that these environments may affect their health, but their health status is overwhelmingly a function of their poverty, age and race.

Social Health

Such a view is implicitly incorporated in the final set of studies we wish to discuss in this section. These stem, too, from the broader definition of mental health to mean peace of mind and general quality of life. This related, as we showed in Chapter 2, to studies of social deprivaton or social *malaise*. In such studies mental illness, conventionally defined as psychiatric disorder, is included as but one factor. An attempt is usually made to include all factors (at least those available in published statistical sources) that can be said to contribute to social and mental well-being, or conversely distress, because in our society low socioeconomic status and minority group membership certainly add to such distress. Indeed, poor social integration (anomie) has been demonstrated as an important component of social ill-being in the United States (Smith, 1973) and Barry, South Wales (Giggs, 1970). Studies of social disorganisation have been carried out for Honolulu (Schmitt, 1966) and Helsinki (Gronholm, 1960). More interestingly, McHarg (1969) has investigated the unity of physical, social and mental health and their identification with specific social and physical environments in Philadelphia. He utilised data on physical disease – heart disease, tuberculosis, diabetes, syphilis, cirrhosis of the liver, amoetic dysentery, bacillary dysentery and salmonellosis – social disease – homicide, suicide, drug addiction, alcoholism, robbery, rape, aggravated assault, juvenile delinquency and infant mortality – and mental disease – measured by both general and child admissions to psychiatric facilities. The three disease-types all reveal a similar pattern, with the highest incidence of disease in the inner city and on the south side and with the lowest incidences in the north-west and north-east. McHarg's work has, however, been taken up by geographers who have used sophisticated statistical techniques. Smith (1973), for example, shows the correspondence of poor quality of life, as measured by housing, crime, health and income, with the black residential districts of Gainesville, Florida and coincidence of deprivations as measured by income, housing, health, education, social disorganisation and participation in the area extending north eastwards from the CBD of Tampa, Florida. In an interesting study of Belfast, Boal *et al.* (1978) utilise three different definitions of social *malaise* ranging from the incidence of a number of problems treated as equally important through a weighting of certain problems to an attempt to get at underlying social processes. The spatial outcomes of the three definitions are not greatly different – social *malaise* is concentrated in the inner city and western sector – but their approach enables them to discuss the societal factors – specifically class and ethnicity – that underlie the patterns. Thus, the localisation of *malaise* is not disputed but its explanation is seen to lie outside the spatial framework.

We can see, therefore, that all our discussions – on physical disease, mental disease and mental health broadly conceived as social *malaise* – end at the same point. Geographical distributions and environmental associations exist. There are localised pockets of disease and ill-health (though related to the size of units of observation), the diffusion of diseases does take a particular spatial course and environmental associations are discernible in relation to both physical and mental diseases, but the causal mechanisms lie elsewhere – in pathogens, the work environment of a group, the poverty, age and culture of a people or the personality of an individual. Even behind the distribution and diffusion of infectious diseases, there lurks the poverty of the people affected. Environment certainly acts as a stimulant to health or disease but explanation lies in economy, society and psychology.

The Impact of Class, Culture and Personality

The differential impact of disease can only be explained in part by environment. Although as geographers we are initially concerned with the distribution of disease and ill-health, the search for explanation must take us beyond this subject's traditional parameters, to look specifically at the totality of life circumstances, at what Dubos (1960, 109) called those 'accidents of birth or life', though we would argue that some such circumstances are less of an accident and more of an allocation, determined by the power structure and belief-systems of the social order concerned. Indeed, beliefs can greatly affect the definitions of health and illness prevalent in a society. Mechanic (1978) argues that 'illness behaviour' – the response to symptoms and the tendency or reluctance to define any symptom as a health problem and to seek medical care – varies between cultural and social groups, while Morris (1975) demonstrates how conceptions of health and illness vary among different groups within one society and between societies, as well as in any one society over time. We refer to our full discussion in Chapters 2, 5, 6 and 7. We should also be aware of the problems of definition and measurement in the field of health and of the doubts cast on the reliability and validity of such data (see Doyal, 1979a; Walters, 1980; DHSS, 1980). Despite these limitations, a picture of social differentiation of disease impact and ill-health does emerge, as does the concomitant differential need for health care (see Chapter 5).

As Brotherston (1976, 73) notes in Britain 'there is so much evidence demonstrating differences in mortality and morbidity between the social classes . . . that it is difficult to select from the evidence'. A recent report (DHSS, 1980) has exhaustively documented these differences. It discovered, for example, that class differences in mortality are a constant feature of the entire human lifetime though they are in general more marked at the start of life and in early adulthood (Figures 3.10 and 3.11). In fact, at birth and during the first month of life the risk of death in the unskilled manual group is double the risk in the professional group. The greatest differences are found in deaths resulting from accidents and respiratory disease, causes of death which are associated with the socioeconomic environment. Between the ages of one and fourteen the mortality rate continues to be closely correlated with class. The greatest differences were found again in deaths caused by accidents (the risk being ten times greater in social class V than I), infective and parasitic diseases and pneumonia. Class disadvantage does become less extreme with increasing age. With adulthood, causes of death become extremely varied and there are significant sex differences in addition to those of class. Thus, circulatory disease and endocrine, nutritional and metabolic diseases of the digestive system affect working-class women most severely and malignant neoplasms, accidents and diseases of the nervous system working-class men. Diseases in which class is significant for both sexes include the infective and parasitic, those of the blood and blood-forming organs, those of the genito-urinary systems and, most important, diseases of the respiratory system.

Although mortality rates have declined over the last 100 years, (see McKeown and Lowe, 1966) there is little evidence to suggest that social class differences in mortality are at present lessening (Brotherston, 1976). Further, Blaxter (1976) points to the deteriorating position of social class V in relation to different disease groups (Figure 3.12). The 'old' diseases – the diseases of poverty – demonstrate a widening gap between the classes while the diseases of affluence – diabetes, coronary disease and appendicitis – are also increasingly concentrated in groups IV and V, possibly showing the differential impact of health education. Walters (1980) also notes the class differences among four diseases traditionally associated with poverty – respiratory tuberculosis, rheumatic heart disease, bronchitis and cancer of the stomach – in 1930–2, 1950 and 1970–2. Men in class V, compared with those in class I, have a far greater chance of dying of one of these diseases. In each successive period there is an increasing difference in the health experience of two classes. Except in the case of rheumatic heart disease death, the differences between the classes are greater in 1970–2 than in earlier periods.

Figure 3.10: Infant Mortality by Sex, Occupational Class and Cause of Death

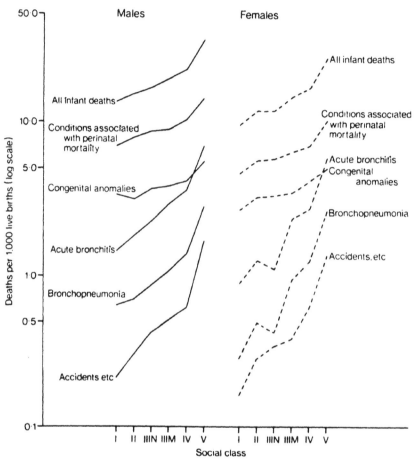

Source: DHSS (1980), p. 35.

Death rates from some other causes are, however, higher in social classes I and II: higher social classes are more prone to death from poliomyelitis, leukemia, cancer of the breast and cirrhosis of the liver. But positive mortality gradients (where rates are high in low social classes) now extend to diabetes, vascular lesions of the nervous system, and coronary disease, and whereas there was no social class trend in deaths from lung cancer and duodenal ulcer in the mid 1930s, twenty later rates were highest in classes IV and V. (Walters, 1980, 122; see also Susser and Watson, 1971; Parker, 1975).

Figure 3.11: Class and Mortality in Childhood (Males and Females 0–14)

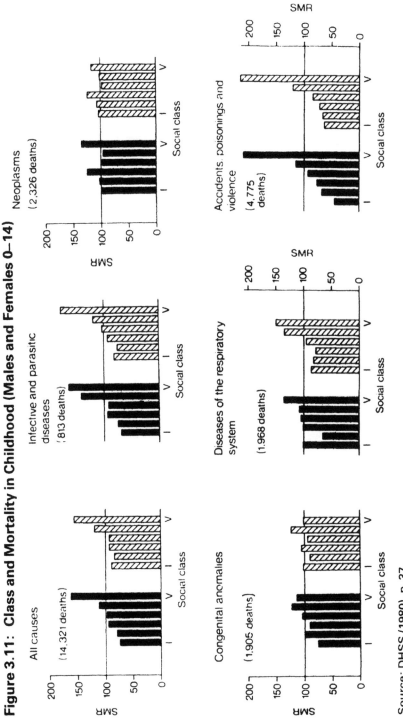

Source: DHSS (1980), p. 37.

Figure 3.12: Mortality Trends for 'Old' and 'New' Diseases by Social Class

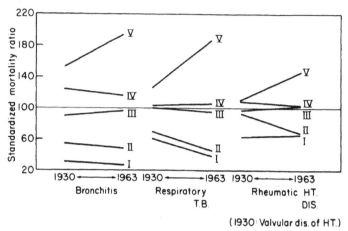

(1930 Valvular dis. of HT.)

Some 'old' diseases: trends of mortality ratios. Men, 1930–63, ages 15–64, England and Wales

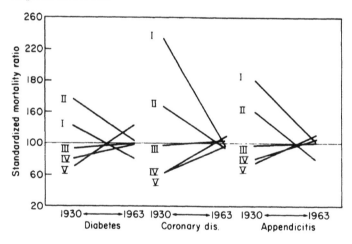

Some 'new' diseases: trends of mortality ratios. Men, 1930–63, ages 15–64, England and Wales

Source: Blaxter (1976), p. 114.

While rates of infant mortality have steadily declined, a similar pattern of class differentiation emerges. In the period 1911 to 1950, the neo-natal and post-neo-natal death rate declined for all social classes in England and Wales (see Morris and Heady, 1955). The difference in rates between class I and V remained almost constant.

For neo-natal deaths, the rate for class V was almost 59 per cent higher than for class I, while in 1949–50 it was 62 per cent higher. For post neo-natal deaths, the figures are 289 and 265 per cent respectively. The post-1950 period has seen a narrowing of class differentials with respect to post neo-natal deaths but these differentials have widened with respect to still-births and neo-natal deaths (DHSS, 1980). Such findings hold for Scotland, too, where decreasing differentials for post-neo-natal deaths have not been matched in overall infant mortality rates (Brotherston, 1976).

These class differences in infant and adult mortality rates are not only found in Britain. Similar differences exist in Finland, Denmark, Norway and France (Tables 3.1, 3.2 and 3.3). Sweden has, however, virtually eliminated social differentiation in mortality rates, suggesting that the reasons for differentiation are to be found not only in the social system but also in the nature of the health care delivery system (see Chapter 6).

Mortality rates do not provide the whole picture of the distribution of ill-health. It is impossible to discover how much ill-health exists; not only does so much depend upon the cultural definitions of health and ill-health but there also exists a complex relationship between feeling ill and clinical assessment which cannot be discovered from hospital or morbidity statistics. The self-reporting of illness possesses a consistent social class variation. Blaxter (1976) notes that class differences occur more in chronic than acute illnesses, with prevalence rates being nearly three times as great in social class V as in class I. Logan and Cushion (1960) point to some interesting differences in consultation rates for a wide variety of groups. With respect to working males, for example, they find above average consultation rates in manual occupations for respiratory disorders, gastric disorders, arthritis/rheumatism and injuries and in non-manual occupations for psychoneurotic and cardiovascular disorders. Overall we find self-reporting and consultation rates increasing for adult males as we go down the social scale. This pattern is, however, reversed after retirement and a more complex profile exists for women. As Brotherston (1976) points out, we do not know how much this gradient is determined by the needs for certification for sickness benefit nor whether the greater frequency of consultation corresponds effectively to the needs of social groups with greater morbidity experience. A comparison of consultation and mortality rates suggests that treatment becomes less adequate with lower social status (Logan

Table 3.1: Denmark: Neo-natal Mortality Rate by Occupation

	1970	1972	1974
Self-employed	10.9	8.1	5.7
Salaried employee	9.9	9.8	7.5
Skilled worker	10.1	8.9	8.1
Unskilled worker	13.5	11.3	9.0
Other/unknown	11.1	10.2	8.8
All	11.0	9.7	8.0

Source: DHSS (1980).

Table 3.2: Finland: Age-adjusted Mortality Indices (1970) by Social Group

		Male	Female
I	Higher admin. or clerical employees, comparable employers and people with academic degrees	78	95
II	Lower admin. or clerical employees and comparable employers	95	100
III	Skilled and specialised workers	92	102
IV	Unskilled workers	148	108
V	Farmers	87	96

Source: DHSS (1980).

Table 3.3: France 1968: Mortality Rates Among Economically Active Men Aged 45–64 (Unstandardised; Per 100,000)

I	Higher cadres (administrators, etc.)	699
II	Industrialists, liberal professions, large commercial proprietors	919
III	Middle cadres (including teachers, medical/social service personnel, army, police)	928
IV	Artisans and small shopkeepers	1225
V	Farmers	1117
VI	Employees (including service workers, clergy)	1392
VII	Qualified workers	1589
VIII	Agricultural workers	1520
IX	Other workers (including miners)	1169
All		1189

Source: DHSS (1980).

and Cushion, 1960; DHSS, 1980), and although the lower groups make more use of doctors in terms of simple numbers, the higher class use them more in relation to the amount of illness they see themselves suffering, and also make about three times as much use of specialists (Blaxter, 1976).

These class variations can also be found in absent from work figures (Table 3.4). The average number of days lost through illness or accident among unskilled manual men was four-and-a-half times that among professional men in 1971 and 1972. Illness also seems to be of greater severity and longer duration among the lower status occupational groups, the retired and unemployed than among those of higher occupational status.

> Rates of long-standing illness rise with falling socio-economic status and tend to be twice as high among unskilled manual males and about 2½ times as high among unskilled manual females as males and females respectively in the professional classes. Inequalities are smaller in childhood and early adulthood and larger in middle age. (DHSS, 1980, 57)

Rates of restricted activity are less clearly differentiated on a class basis. Specific illnesses are, however, so differentiated – bronchitis (College of General Practitioners, 1961), childhood infective diseases (Walters, 1980), poor dental health (Blaxter, 1976).

Class variations can also be discerned in the distribution of psychiatric disorders. Although the associations between social class and mental disorder are variable and inconsistent in children, the situation is quite different in adults. The great majority of studies have shown that the highest rates of psychiatric disorder are in the lowest social groups (Dohrenwend and Dohrenwend, 1969). Rutter and Madge (1976) point out that important qualifications have to be made to this statement. In hospital-based studies, for example, associations with low social status are most marked for schizophrenia and least marked for neurosis and depression. Studies in the general population have shown that depressive or neurotic disorders are commoner in working-class women (though not men) and that these women are less likely to seek medical help than middle-class women. The relationship between class and disorder is also not straightforward. Thus, schizophrenia leads to social deterioration rather than low social class leading to schizophrenia (see Goldberg and Morrison, 1963; Wardle, 1962).

Table 3.4: England and Wales: Working Males Absent from Work Due to Illness or Injury

Socio-economic group	Absent from work due to illness or injury in a two week reference period – rate per 100			Average number of work days lost per person per year	
	1971	1972	1977[a]	1971	1972
Professional	37	21	20	3.9	3.1
Employers and managers	37	39	20	7.2	6.2
Intermediate and junior non-manual	44	48	50	7.6	6.0
Skilled manual	57	56	60	9.3	9.4
Semi-skilled manual	56	68	70	11.5	10.5
Unskilled manual	88	99	60	18.4	17.6
All groups	52	54	40	9.1	8.4

Note:
a. Rate given only to nearest 10.
Source: DHSS (1980).

This idea of an illness-determined downward drift lends support to the drift hypothesis concerning the location of schizophrenics in the city discussed earlier. Further, little is known on a general level about the association between class and disorder. 'The social class differences could be the end-product of an intergenerational accumulation of genetically vulnerable individuals or they could represent the environmental effect of social factors associated with occupational status' (Rutter and Madge, 1976, 214). Indeed, Brown *et al.* (1975) have showed that working-class women with young children experienced far more acute and chronic stresses than did similarly placed middle-class women. But the proportion with stresses did not differ by social class in the case of childless women or those without children at home.

Thus, class, like environment, is not the only factor in explaining psychiatric disorder; stage in the lifecycle becomes important. Brown *et al.* (1975) found that the presence of three or more children at home predisposed women to depression. In fact, family size and disorder seem to be closely related. With psychiatric disorder, however, there seem to be predisposing factors which may create problems if the personality of the individual is of a specific type. Thus, family circumstances, patterns of upbringing and socialisation, environment and class interact with genetic and biological factors to present a multi-causal framework for understanding and explaining mental health and ill-health.

It is possible to say much the same about physical health and ill-health. Although class seems to be the overriding differentiating factor, there are notable age, sex and racial differences. Thus, for example, the mortality rates of males at every age are higher than of females and in recent decades the differences between the sexes has become relatively greater. Men always have a higher incidence of cancer (other than of the reproductive organs) than women. Thus, for primary liver cancer in Singapore the ratio men to women is 7 to 1 and in Liverpool 5 to 1. An explanation must account for the higher incidence in both the industrialised and non-industrialised world and in cultures that both readily accept alcohol and prevent its imbibation. It is, in fact, argued that the presence of testosterone helps induce hepatic tumours (see Roberts, 1976). It has also been discovered that the pattern of physical illnesses in blacks differs from that in whites. Thus, Asians may have brought their higher rates of tuberculosis from its endemic areas (Thomas, 1968). Oppé (1964) notes that Jamaican women are particularly prone to rubella because of their lack of natural immunisation while West Indian children suffer disproportionately from respiratory infections perhaps aggravated by poor living conditions. Indeed, it is possibly true to say that the incidence of physical illness among ethnic minorities is greatly influenced by the physical conditions in which they live and bring up their children – a matter then of environment, the type of which is itself determined by income and occupation, that is class. Biology and economy seem, therefore, to be major causes of variations in illness and disease incidence.

Conclusion

The relationship between man, disease and environment is far from clearcut. The utility of the traditional medical geography can be seen at the macrospatial scale and in the demonstration of environmental associations and diffusion processes. Thus, Howe (1976, 56) indicates that:

> cancer of the liver and mouth are more frequent in southern Africa and India than in Europe or North America. The reverse

is true for cancer of the large intestine. Cancer of the lung and bronchus appears to be more common in Britain than anywhere else in the world, though it is also widespread in central and eastern Europe, USSR, USA. Incidence is slightly less in Canada, Australia and among the white population of South Africa and considerably less in Scandinavia, South America, Asia.

The highest incidence of stomach cancer is in Japan and Soviet Asia – but its incidence has declined in recent years – Europe and North America, leading to the suggestion that carcinogenic agents could be produced by the deterioration of foodstuffs, combatted by food preservation and refrigeration. High incidence of oesophageal cancer in Brittany, Curacao, Jamaica and Malawi is associated with local maize beer or porridge. Howe also points, for example, to malaria and yellow fever and their association with the mosquito and how diet, housing construction and treatment of domestic animals can affect their diffusion (Figures 3.13 and 3.14). Howe rightly suggests that such distributions reflect ecological imbalances in different parts of the world, but in so saying demonstrates both the strength and weakness of medical geography. It does point to ecological associations and to the variations in distribution, but such factors cannot explain why certain individuals and groups suffer more than others from disease and ill-health. This problem is worsened when we focus attention on the general health status of the population rather than on specific diseases. To explain the differential social impact of ecological associations and the varied nature of health status and experience we must consider the broader factors of society and economy, with due allowance for genetic and biological forces.

Thus, for example, it is important to note the non-random nature of accidents as a collective class of events (DHSS, 1980), with boys in social class V being five times as likely to die before reaching school age than those in class I. Both economic and cultural factors are implicated – the material resources of lower status households are such that children are more likely to be left to their own devices, a state of affairs compounded by this group's norms of adventure and daring. Birch and Gussow (1970) indeed suggest that an unsatisfactory social environment can be a cumulative hazard which is likely to be worsened by inadequate nutrition, resulting in slow physical growth and possible disruption of brain development.

Figure 3.13: World Distribution of Malaria

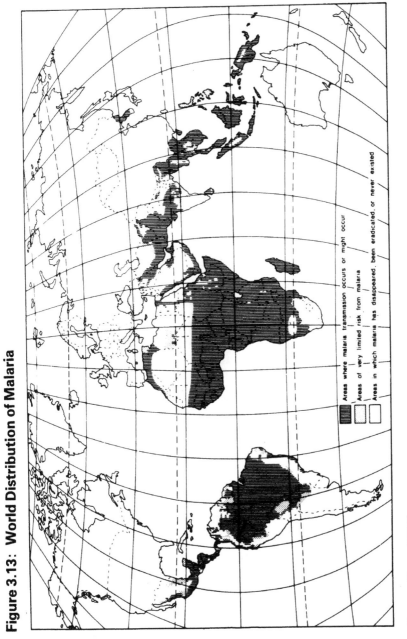

Areas where malaria transmission occurs or might occur

Areas of very limited risk from malaria

Areas in which malaria has disappeared, been eradicated, or never existed

Source: Howe (1976).

Figure 3.14: Yellow Fever, Endemic Zones, 1973

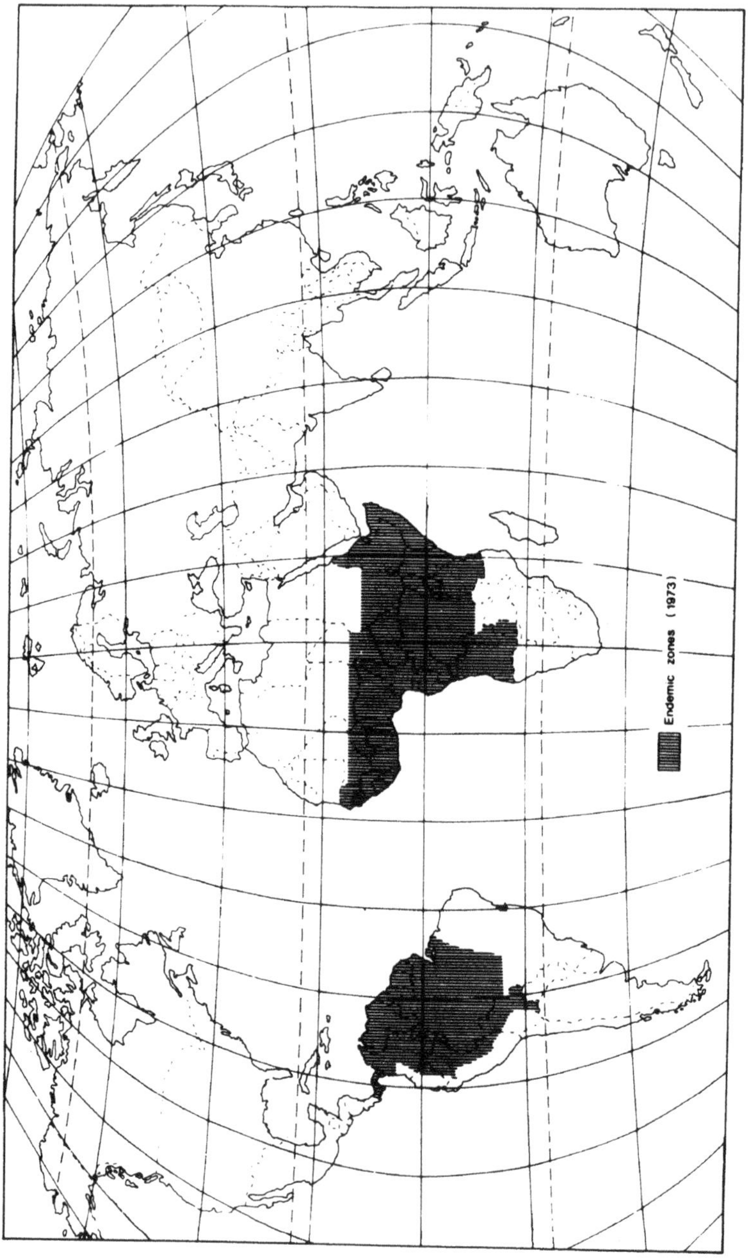

Endemic zones (1973)

Source: Howe (1976).

We have already noted the class differences in mortality and morbidity in adult life. Individual factors do play a part in, for example, the adoption of a healthy lifestyle – such as rejection of smoking, emphasis on diet, exercise, etc. But again the predisposition to undertake such a lifestyle is class-related and systemically constrained (see Chapter 2). And in terms of liability to accident, disability, long-term illness, occupational disease and death, the manual worker is at by far a greater risk than the non-manual worker. Wear and tear – without adequate compensation – on body and mind have greater impacts on manual workers. 'Ageing' is, therefore, a process that occurs sooner, rather than later, for male manual workers.

The differential impact of disease and ill-health is thus inextricably linked, through class, to the operation of the economic system. Brenner (1979), for example, suggests that rates of mortality are linked to national economic performance, specifically that the rates of some causes of death rise in times of unemployment and economic crisis. In a response to his critics (see particularly Gravelle *et al.*, 1981), he has found a positive relation at a zero or one year time lag between unemployment and all causes of death except for infant mortality, where no relation is found in Scotland, and suicide, where the relation is found only for Scotland and at a two-year lag (Brenner, 1981).

Such findings seem to posit a materialist explanation for levels of health (see Chapter 1).

> In the advanced capitalist societies, surplus value, i.e. the excessive extraction or exhaustion of human bodily resources and effort, in the production process, is realised not through the depression of real wages . . . but through hazardous, punishing and physically stressful work processes; human immiseration is no longer manifested in terms of grinding poverty and deprivation, but finds expression in the spiritual and intellectual impoverishment of industrial workers . . . It is in the actual process of commodity production, therefore, that disease is produced, through physical stress engendered by tense competitive work relations and routines and through social stress manifested in neglected or disrupted networks and relationships in the realm of domestic and community life. (DHSS, 1980, 164)

Further, social reproduction demands pollution and the

consumption of commodities hazardous to health. We would contend that this is a powerful explanation of health status inequalities. By itself, it is too sweeping. Environmental, individual and cultural factors do appear significant in explaining some of the variation in inequalities. And they are also significant for our understanding of the development of health care systems (see Chapter 6). The materialist explanation does point to the increasing rationality of production and the impact of such rationality on workers: a rationality that is extended to the planning of ameliorative facilities (see Chapter 7). But it is true to say that the policy initiatives and planning formulations that we are about to examine are, on a conscious level at least, shaped more by the multi-causal explanation of ill-health and the health status indicators on which such explanation is based.

4 PERSPECTIVES ON THE LOCATION AND DISTRIBUTION OF HEALTH SERVICES

Introduction

The most tangible features of health care delivery systems are the clinics, surgeries and hospitals to which we usually turn in times of illness. Despite the clear evidence of earlier chapters which illustrated the role of the environment in generating ill-health, societal effort has (in the Western and developed worlds at least) been directed into the provision of systems of clinical services intended to combat episodes of illness rather than into environmental modification for the specific purpose of improving health. There are, of course, exceptions: slum clearance, the development of town planning, the introduction of air pollution controls and occupational health legislation are all examples of environmental modification undertaken with some health objective in mind. Leaving aside the question of whether environmental improvement achieves more than conventional medical services, it is the provision of such services which still constitutes society's most obvious response and investment in health with the implicity, though increasingly challenged notion, that more of such services are desirable and will improve health. On occasion this viewpoint has been made explicit; in the context of the NHS, Sir Francis Avery Jones, critising proposed changes in the distribution of medical services, wrote:

> Within the main cities there are large areas with serious problems of urban deprivation; overcrowding; incomplete domestic services; unemployment; and increased mobility of population with resulting lack of friends and relations, dreary streets and traffic pollution. Additional resources and *more hospital beds* are needed to meet the medico-social needs of such areas. (Avery Jones, 1976, 1048; emphasis added.)

Ultimately we question this belief, but in this chapter we examine the perspective adopted by geographers and others to the location and distribution of such services. Implicit in these approaches is the belief that health needs can be quantified, and the distribution of services optimised with respect to such criteria as distance/cost minimisation or accessibility maximisation, in the context of a concern for distributional equity. We emphasise that these perspectives are often theoretical and basically 'economic' (see Chapter 1), and we argue in the next chapter that the real world is more complex as a consequence of various constraints. In our last section we examine a practical attempt to implement a policy designed to achieve an equitable distribution of financial resources devoted to health care provision. This policy – devised by a body known as The Resource Allocation Working Party (RAWP) (Gt Britain, 1976b) – set out to quantify the resource needs of NHS administrative units at a variety of geographical scales, and thence to allocate resources in proportion to those needs. In keeping with the rest of the chapter we limit our critique of the policy to the technical difficulties of measuring and quantifying, adopting a broader critical perspective in Chapter 7 after we have explored the effects of various constraints (see Chapter 5), enabling us then to set the policy and other perspectives developed here in a wider socio-political context. We commence, however, with one of the geographers traditional concerns – the impact of space on service utilisation.

The Impact of Location on Facility Utilisation

Throughout this chapter we recognise the impact which discrete locations have on service utilisation, so before proceeding to examinations of the locational efficiency of health care delivery systems it is appropriate to introduce a number of studies which have explored the effect of distance on provider-patient contacts. Distance decay effects are familiar to geographers in the context of retailing and as an element of the gravity models developed by Stewart (1948) and Zipf (1949) from the pioneering work of Ravenstein. Shannon and Dever (1974) suggest, however, that distance decay effects had been noticed somewhat earlier in the context of health care. They draw our attention to Jarvis' Law, formulated on the basis of observed mental hospital use, which states: 'The people in the vicinity of lunatic hospitals send more patients to them than those of a greater distance' (quoted in Shannon and Dever, 1974, 111). Lower admission rates at increasing distance from the mental hospital are attributed to greater knowledge of its beneficial potential amongst nearby residents rather than by spatial variation in psychiatric morbidity. This line of argument has guided empirical studies which have attempted to quantify the friction of distance. Usually this is achieved by calculating facility attendance or utilisation rates of people distributed at various distances around the facility. The utilisation rate is adopted as the dependent variable and distance from the facility to place of residence as the independent variable. Calculation of the regression coefficients generates the gradient of the regression line, which indicates the severity of the distance constraint. Steep negative regression slopes indicate greater friction, that is utilisation rates decrease rapidly as distance from residence to facility increases.

One of the most interesting examples of this approach is provided by Jolly and King (1966) who examined the effect of distance on patients attending aid posts, dispensaries and hospitals in part of Uganda (see Smith, 1977). Figure 4.1 shows the basic distance–attendance relationship for hospital outpatients and its conversion to a log-linear relationship when the dependent variable is plotted on a log-scale. The regression line for each of the services is plotted and the frictional effect of distance can be observed. Some people are prepared to travel up to 20 miles for the services of a hospital, but the maximum distance travelled to an aid post is no more than nine miles. Significantly, in both cases there is a sharp decline in the average number of attendances as distance increases. In the case of the aid post, attendances halve for every extra mile to be travelled; the effect is less in the case of hospitals, halving every two miles. The relationship between the type of service and the effects of distance on utilisation has important planning implications. In low order services like the aid post with large distance decay coefficients it may be more sensible to take the service to the people if the objective is to maximise the distribution of care.

The relationship between distance and utilisation has also been examined in Walmsley's (1978) study of a rural hospital in New South Wales, and Ingram *et al.*'s (1978) study of attendance at a hospital emergency department in Toronto, Ontario. Walmsley investigated the effect of distance on both outpatient and inpatient attendance rates per 100 resident population, using the shortest road route between patient residence and hospital as the measure of

Figure 4.1:   Distance Decay Effects on Service Utilisation

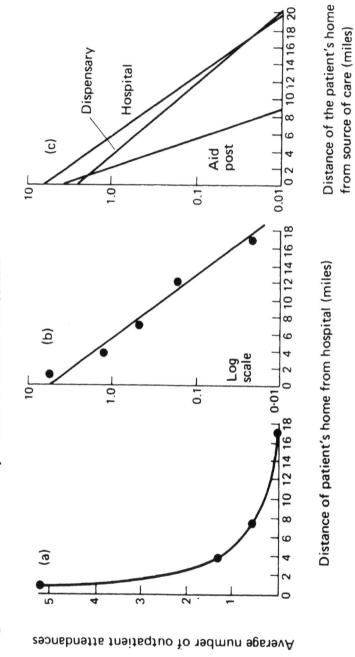

Source: After Smith (1977).

Figure 4.2: *Per Capita* **Usage of Humber Memorial Hospital's Emergency Department (Population Data for 1971)**

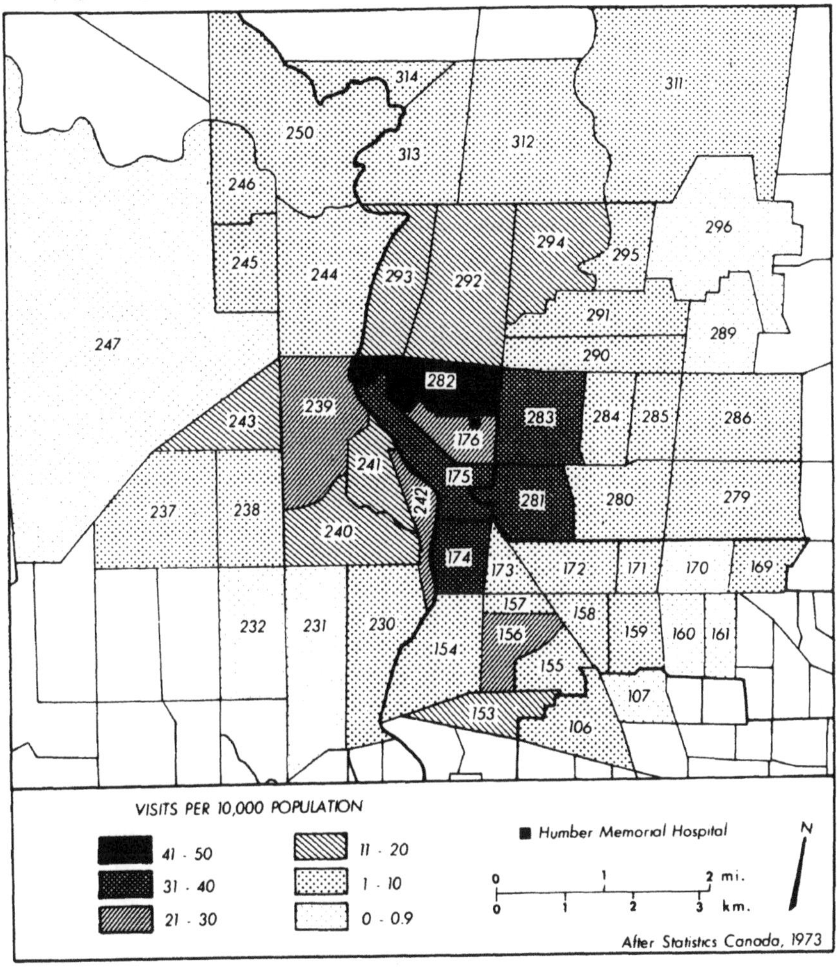

Source: Ingram *et al.* (1978), p. 60

distance. Significant negative correlation coefficients were derived confirming that hospital use by inpatients and outpatients declined with increasing distance from hospital. Ingram *et al.* found in their study the existence of a similar relationship, and were able to show that 67.5 per cent of the variation in inpatient visits per 1,000 population was accounted for by linear distance (Figure 4.2).

In an evaluation of community hospital policy in East Anglia, Haynes and Bentham (1979) also highlighted the impact of distance on utilisation. They investigated its impact on outpatient attendances by comparing the actual attendance rate as reported in surveys with the rate expected (derived from age/sex specific national data) for populations living at various distances from King's Lynn, the town where outpatient services were located. Table 4.1 presents the results of this analysis and shows clearly that those living more than ten miles from King's Lynn made 20 per cent fewer outpatient visits than their age/sex characteristics suggest they should. This study also showed the same general relationship with regard to visitors, especially visitors to preconvalescent patients whose visits declined with increasing hospital–home distance, independently of length of patient stay, another factor which reduced the number of visits.

Table 4.1: Actual and Expected Visits of Outpatients by Distance from King's Lynn

Distance from King's Lynn	Actual %	Expected %	Ratio of actual expected
Within King's Lynn	45	27	1.67
< 10 miles from King's Lynn	26	29	1.08
> 10 miles from King's Lynn	29	49	0.59

Source: Haynes and Bentham (1979), p. 125.

It is important to point out at this stage that distance is but one, albeit important, determinant of access. As Phillips (1981) emphasises numerical frequency of attendance does not capture important differences in the need to use a service, and in his behavioural study of spatial patterns of general practitioner (GP) attendances he examines the impact upon them of social status, personal mobility, age structure and previous residence. Although we return to this theme in the next chapter where we point to important social constraints on service utilisation, it is useful to introduce this perspective at this point lest we assign too much significance to distance as a determinant of use. Phillips' study was undertaken in West Glamorgan, Wales and aimed to investigate the availability and use of GP facilities by populations in identified socially distinct sub-areas (based on 1971 Census Enumeration District (ED) data). Three pairs of EDs, each consisting of one low status and one high status ED, were selected with similar access to

GP services. In addition, a fourth pair of EDs (both low status) were identified, but one had local GP services and the other distant GP services. Within this framework Phillips was then able to evaluate the impact of distance and the socio-economic variables referred to above. The results of Phillips' analysis emphasise how these variables interact; the pattern of use, whilst having some relationship to distance, was by no means explained entirely by it. Of the factors which Phillips (1981, 141) investigated 'social status, personal mobility and previous residence to a greater or lesser degree, appear to influence the surgery to attend'. Of these, Phillips (1981, 141) singled out 'place of previous residence' to be 'highly significant in explaining spatial patterns of utilisation behaviour'. In other words, there is an important element of historical inertia in patient use of GPs; once chosen, and presumably satisfied with the service, patients, if they are able to overcome the intervening distance, continue to use services located some way from their current homes.

None the less, denial of access as a consequence of location can have an impact on health. In Newcastle-upon-Tyne, Bradley *et al.* (1978) investigated the relationship between distance–decay and dental decay in school children. The authors were able to assess the independently related variables of services, access and social class on dental health by calculating partial correlation coefficients. These showed the spatial access variable to be predominant. Whilst it is generally argued that dental health education leads to improved dental health, Bradley *et al.* assert that the relocation of dental services could have a more immediate beneficial effect. Given then that facility location does have an impact on health service use, we turn our attention to the way in which geographers have approached the question of distributing and locating health care services.

Locational Efficiency

In classical industrial location theory the optimum location for a firm is identified in relation to spatially variable costs and profits. However, unlike individual firms and industrial enterprises medical services are commonly non-profit seeking institutions whose locational position will not be governed by the need to generate profit either by cost minimisation or revenue maximisation, though the degree to which this operates depends in part on the nature of the containing society (see Chapter 6). Instead, the objective is to maximise some benefit or minimise some cost to society as a whole and this may be unquantifiable in monetary terms (Revelle *et al.*, 1970). The location problem of medical services further differs from the classical location problem because individual medical facilities are usually part of an integrated system of services, so that the location problem can only be properly conceived in that context (Teitz, 1968). Within these health care delivery systems there are different types of service operating at varying ranges and population thresholds, from the individual doctor in his surgery to the teaching hospital complex with its numerous specialist departments. Our problem then is that of how to arrange in space a hierarchically ordered system of medical facilities which minimises some social cost or maximises a benefit.

Figure 4.3 provides one solution by using the simplifying assumptions of central place theory. In this scheme the objective is to locate a medical system consisting of three tiers (general practitioner, health centre and general public hospital) on a uniform plain with an even population distribution such that, subject to considerations of range and threshold, the facilities within each tier are optimally located. The result is the familiar nested hierarchy of hexagonal service areas. Although it is unlikely that such an ideal solution will ever be found in reality, the model does capture some of the basic elements of the location problem posed by health services. In the geographical literature there are a number of studies which have set out from this point to determine the optimal distribution and specific location for health care facilities at different levels of this functional hierarchy.

Primary Care: General Practitioner Services

One of the most common practices adopted by geographers at the primary care level has been to consider the spatial distribution of general practitioners. In answering the question of where general practitioners should be, geographers have attempted to identify the deficiencies of existing distributions. They have not in general attempted to identify the precise location for such practitioners, but have used practitioner–population ratios to identify doctor-rich and doctor-poor areas. Underlying these analyses is the concept of proportional equality, whereby each territorial or administrative unit should have the same number of physicians in proportion to the

Figure 4.3: Central Place Hierarchy of Medical Facilities

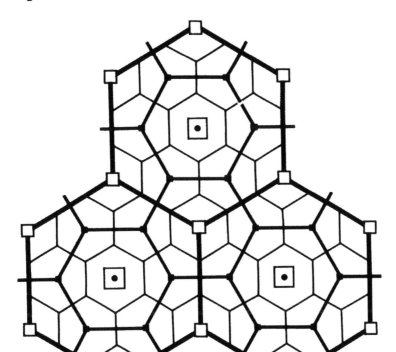

■ **REGIONAL HOSPITAL (>500 beds)**

□ **DISTRICT HOSPITAL (101-500 beds)**

● **RURAL HOSPITAL (20-100 beds)**

Source: Shannon and Dever (1974).

size of its population. As Smith (1977) has argued, equality conceived in this way is meaningless if there are differential needs between populations which are not captured by population size alone. In fact, an equitable distribution may only be achieved by a very unequal geographical distribution, if that distribution is in accord with some criterion of need (see below). Even so, geographers have been content to produce a variety of studies at different geographical scales – international, regional and intra-urban – and in different political contexts – which demonstrate the existence of unequal and assumed inequitable distributions of general practitioners by reference to doctor:patient ratios.

In the United States, Shannon and Dever (1974) present data relating to 1970 which shows the north-east of the country to have a general practitioner:person ratio of 0.66 per 1,000 people, compared with a national average of 0.53 per 1,000, and a ratio of 0.44 per 1,000 in the south (Figure 4.4). The implication is that the south is unjustly treated whilst the north-east is well off for general practitioners. Such differences are not confined to the Western developed world. Cole and Harrison (1978), Ryan (1978) and Kaser (1976) all show how in the Soviet Union regional differences in the number of physicians (including general practitioners) *per capita* still persist even though their total number has increased considerably.

Some of the largest differences exist between rural and urban areas – especially in under-developed nations. Potts (1976) presents data to show how doctors in such countries are concentrated into capital cities. In Kenya, for instance, he reports a doctor:patient ratio of 1:672 in the capital, whereas in the remainder of the country the equivalent ratio is 1:25,000. Differences of this magnitude do not generally occur in the urban areas of developed nations, but a large number of studies have drawn attention to significant variations in the availability of medical personnel. Shannon and Dever (1974), de Visé (1971) and D.M. Smith (1979) all document the magnitude of differences in American cities and Stimson (1977, 1980) draws attention to similar variations in Adelaide, Australia. In this latter city doctor:patient ratios varied between 0.34 per 1,000 people in the suburb of Woodville, to 1.38 per 1,000 in Glenelg. As in the United States these differences have a clear relationship to the social geography of the city with, in general terms, higher status suburbs having better, that is higher, doctor-patient ratios. In these countries medical practitioners derive their income from private practice and so there is an inevitable attraction to wealthier suburbs. None the less in a country like Britain, which has a state financed and regulated general practitioner service, there are still considerable variations in doctor:patient ratios related to social geography. At a regional scale Coates and Rawstron (1971) showed how the south-east of England had more general practitioners *per*

Figure 4.4: Distribution of General Practitioners in the USA, 1970

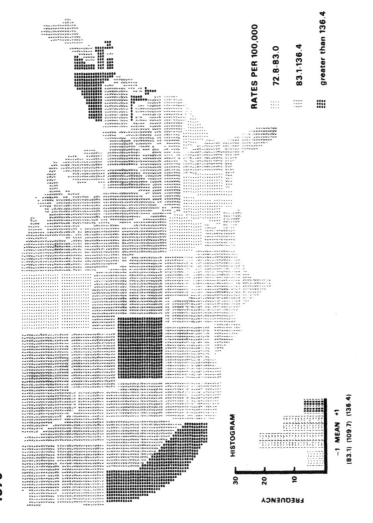

Source: Shannon and Dever (1974).

Figure 4.5: Distribution of Hospital Beds in the USA, 1970

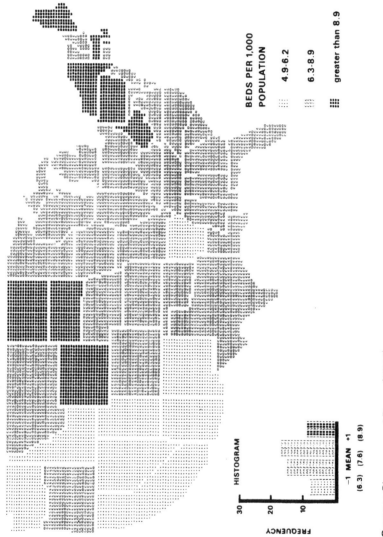

Source: Shannon and Dever (1974).

capita than the north and west and that within these regions, lower status towns and suburbs faired worst of all. The persistence of these inequalities is significant given the efforts which have been made to regulate the distribution of practitioners working in the NHS. A Medical Practices Committee exists to control the appointment of practitioners with the declared objective of achieving a more equitable distribution by defining areas of the country as 'restricted', 'intermediate', 'open' and 'designated' according to the average patient list size of existing practitioners in the area. List size thus becomes the criterion of need. Restricted areas where the average patient list size is less than 1,800 are closed to new practitioners, whereas the designated areas (average list size greater than 2,500) are open to new practitioners who may also receive a cash bonus to establish a practice (Butler *et al.*, 1973).

Despite these measures, spatial discrepancies in doctor:patient ratios have prompted a number of studies which have taken inequality of access as their theme. At an intra-urban scale Knox (1978; 1979) has developed this theme by investigating the relationship between the distribution of practitioners and the distribution of social deprivaton. In a study of four Scottish cities (Aberdeen, Dundee, Glasgow and Edinburgh) Knox (1978) related the practitioner distribution to socioeconomic characteristics of the population disaggregated into sub-areas of the city. From this analysis Knox reached three general conclusions: few surgeries were located in the CBDs of the cities, though many were located adjacent to them; few surgeries were located in post-war peripheral estates (including owner-occupied developments); and there was a relative concentration of surgeries in poorer inner city areas. Knox advances the notion of historical inertia to explain these patterns, that is the surgeries have remained close to the city centre whilst the population has moved out. This is a pattern common to most British cities, and has been referred to elsewhere (Phillips, 1981). Proximity to a surgery, however, does not guarantee access to care, since the distribution, density and mobility of residents must also be taken into account. Knox achieves this by constructing an index of accessibility using the familiar gravity model, but weighted in such a way that car ownership and travel speeds (by both private and public transport) are taken into account. Variations were observed from city to city, but Knox reached the general conclusion that lower status areas, especially peripheral local authority housing estates, were especially disadvantaged in their access to general practitioners. Problems of access are thus added to the other social problems faced by residents in such areas.

Knox (1979) takes this point up more fully in a second paper which extends his considerations to an analysis of the relationship between medical deprivation (availability of general practitioners) and area deprivation. The general conclusion reached is that the two are spatially coincident confirming what Hart (1971) has termed the 'inverse care law' whereby the availability of good medical care varies inversely with the need for it. These differences in the availability of services do not occur by chance. They are the spatial manifestation of particular social processes and forces combined with various historical factors or, as we have chosen to call them, constraints, which lead to deviations from an even, and supposedly spatially equitable, pattern of services.

Secondary Care: Hospital Services

The term hospital services hides a multitude of hospital types, including celebrated teaching hospitals with distinguished medical staffs, through more humble, small local hospitals to specialist institutions dealing with particular forms of illness, the most obvious example of which are psychiatric hospitals. Although geographers have adopted methods applicable to all of these services, most studies have concentrated on general hospitals providing acute inpatient and outpatient care. As in the case of general practitioner services, a body of literature highlighting variations in hospital bed:population ratios has developed. Parallel with this is another set of studies directly derived from classical location theory and concerned with the identification of precise locations for the siting of hospitals. In both instances the objective is to improve the distribution of services by, on the one hand, increasing the volume of services in particular localities and, on the other hand, improving their accessibility to potential patients. The 1962 Hospital Plan for England and Wales (Gt Britain, 1962) succinctly states the framework within which these studies are based. When describing the hospital system inherited by the NHS in 1948 it noted:

> The hospital system was largely the product of historical causes. Changing notions of the right place for a hospital meant some of them were in the wrong place by modern standards; others once

conveniently situated were no longer so because of movements of the population (Gt Britain, 1962).

Consequently in the UK it is easy to point to regional and intra-urban differences in the level of general hospital services (Table 4.2) with the south-east of England being particularly well supplied, especially inner London. Likewise, many of the studies of general practitioner services referred to earlier also explored the distribution of hospital facilities as reflected in bed:population ratios, and they too have reported similar geographical inequalities (Figure 4.5) (Shannon and Dever, 1974; Stimson, 1977; 1980; Cole and Harrison, 1978; Smith, 1977).

Table 4.2: Regional Variations in the Availability of NHS Hospital Beds, 1978[a]

	All special-ties	Acute special-ties[c]	Medical	Surgical	Mental illness	Mental handi-cap	Obste-trics[b]
England[d]	8.1	2.9	1.1	1.7	2.0	1.1	2.2
Northern	8.4	3.2	1.2	1.9	2.0	1.1	2.3
Yorkshire	8.6	3.1	1.2	1.8	2.2	1.0	2.3
Trent	7.1	2.4	0.8	1.4	1.7	1.1	2.1
East Anglian	7.2	2.5	0.9	1.5	1.8	0.9	2.1
NW Thames	8.5	3.1	1.3	1.8	2.5	1.2	1.9
NE Thames	8.4	3.5	1.6	1.9	2.0	0.8	2.3
SE Thames	8.4	3.2	1.3	1.9	2.1	1.0	2.0
SW Thames	9.5	2.6	0.9	1.6	2.9	2.2	1.9
Wessex	7.3	2.5	0.9	1.5	1.8	0.9	2.2
Oxford	6.3	2.4	1.0	1.4	1.2	0.9	2.0
South-western	8.4	2.4	0.8	1.6	2.0	1.7	2.0
West Midlands	7.2	2.6	1.0	1.6	1.7	1.0	2.1
Mersey	9.3	3.5	1.3	2.0	2.7	1.0	2.3
North-western	7.8	3.2	1.2	1.9	1.6	1.1	2.3

Notes:
a. All figures (except Obstetrics) relate to averge daily available beds per 1,000 population.
b. Obstetrics figures are average daily beds per 1,000 female population aged 15-44.
c. Acute specialties include all medical specialties, all surgical specialties, gynaecology and pre-convalescent department.
d. England figures include London Postgraduate Hospitals which are not included in the regional figures.
Source: Derived from Great Britain (1980), Table 4.11, *Health Personal Social Service Statistics for England, 1978.*

In addition, however, many geographical studies have approached the distribution of hospitals from the standpoint of locational analysis, attempting to identify the optimum locations for services and the utilisation consequence of discrete locations. Borrowing directly from Weberian industrial location theory Smith (1977) describes how the optimum location for a maternity hospital in the city of Sydney was identified. The objective in the example was to locate a single new maternity hospital so as to minimise the total aggregate travel of patients, and Smith used a modification of the Kuhn and Keunne algorithm (fully described by Taylor, 1976) to determine this point. The result was the identification of a location seven miles west of the city centre where hospital services are currently concentrated.

A much more complex problem is the simultaneous location of a system of facilities, a requirement when services are provided through multiple outlets. These outlets need not be complete hospitals; they may be individual departments to be added to an existing structure – a casualty department or a specialist kidney unit for instance. One of the earliest examples of this kind of problem is described by Godlund (1961) who was attempting to identify sites for regional hospitals in Sweden. Using isodapane (equal cost) and isochrone (equal time) maps in conjunction with maps of population density Godlund identified seven centres for regional hospitals which maximised the proportion of the population living within four hours travelling time and which also minimised the aggregate travel time and cost for the total population.

Since Godlund undertook this analysis the explosion of computer technology has led to the proliferation of computer algorithms which find solutions to what have become known as location–allocation problems. Linear programming methods have been commonly employed; Cox (1965) has described their application, and Rushton et al. (1973) have fully documented the variety of algorithms available. Abler et al. (1971) provide one of the most accessible presentations of these methods with an example based on Gould and Leinbach's (1966) attempt to locate a system of three hospitals in Western Guatemala. The attraction of these methods lies in the relative ease with which alternative systems can be evaluated against the identified optimum, the difference being regarded as a measure of locational inefficiency, and it was essentially this kind of approach which was employed in the Chicago Regional Hospital Study (Morrill and Kelley, 1970;

Morrill and Earickson, 1969).

Commonly the objective of location–allocation models is the identification of a series of locations which minimise the total aggregate travel of potential patients, who are then assigned to particular hospitals. Travel cost is usually based on some variant of linear distance, though there is no reason other than practical why measures of financial cost or time should not be employed. The optimal solutions identified by these models are usually identified as 'efficient' solutions, though as Symons (1971) and Schneider and Symons (1971) argue, they are not necessarily the most equitable solutions, in the sense that cost minimisation is achieved by requiring some members of the population to travel further than others. In short, maximum average accessibility for the whole population is obtained by an unequal distribution of travel distances amongst the population and this may be considered unjust. This point was taken up most fully by Schneider and Symons (1971) who provide a framework within which the conflicting considerations of equity and efficiency can be evaluated. A distinction can be made between 'efficiency', 'equity' and 'welfare' levels of access (a point we shall address in broader terms in our discussion of social justice in Chapter 6). Efficiency is measured by the minimum mean average travel distance, equity the minimum standard deviation of travel distances and the welfare solution lies at some point between these (see Smith, 1977).

There is, of course, no objective rule for resolving the conflict of equity and efficiency. Ultimately political and administrative practice will determine the solution to be adopted. Moreover, these perspectives on the location of medical facilities deliberately exclude non-spatial factors as the determinants of location and utilisation of services. We believe that these non-spatial factors are probably more important than distance alone, but we believe distance is an important factor generally underestimated by planning authorities. To concentrate on distance to the exclusion of other factors may be regarded as a sterile exercise, but at the same time it is easy to overlook the importance of physical proximity. In the case of medical emergencies physical proximity is vitally important, but in non-urgent cases it is not an issue which has captured the imagination of medical care planners who, often people with high mobilities themselves, neglect the significance of distance to those with lower mobility. The neglect of distance as a factor in utilisation behaviour may be a valid perspective when major non-spatial barriers to care exist – such as race, income, or religion as in the USA – but when such forces operate to a lesser extent, as in the NHS for example, physical proximity can become a pre-eminent cost in the potential patient's decision to travel. Some of the opposition which health authorities have met when closing hospitals down can be attributed in part to the realisation by potential patients that closure transfers costs to them. Savings from scale economies which accrue to health authorities are achieved by transferring additional transport costs to patients and visitors. As Teitz (1968, 44) observes: 'locations of outlets for "free" public services are not easily moved in the face of opposition from citizens who know they are not free'.

Discussing these issues in the Soviet context, D.M. Smith (1979) highlights the explicit recognition given to them when he reports an informant explaining that the decision to centralise services is a question of whether to save time for the people or money for the state. Even though we recognise that patient travel distance can only be one consideration in the decision of where to locate a complex hospital, modelling techniques such as those we have described do at least provide a basis with which to compare and evaluate administrative practice.

Spatial Resource Allocation Policies in the NHS

In this section we widen the issue of distributional equity by examining the development of a policy with the specific objective of allocating financial resources in proportion to measured medical need. We are not concerned here with the problem of discrete facility locations. Rather our concern is with the allocation of financial resources to particular places to ensure that people at equal risk have equal access to medical care in the sense that variations in resource levels are justified by reference to a number of criteria of need (see below). The context in which these policies have emerged is geographical inequality in the distribution of health services, with a concentration of medical facilities and manpower in south-east England. The pattern of hospital provision today is largely the product of history; it is a reflection of nineteenth-century population distribution coupled with unco-ordinated philanthropy, the effects of which are demonstrated by the number of teaching hospitals within inner London (see Chapter

5). As Abel-Smith (1964, 405) put it, 'the pattern of [hospital] provision depended upon the donations of the living and the legacies of the dead rather than on any ascertained need for services'.

Although the NHS was intended to achieve equality of access to health care for populations at equal risk, resource allocations between 1948 and 1975 effected little change to the historical pattern. They were incrementalist rather than redistributive and, in Brotherston's words, involved the 'use of last year's budget with a bit added here and bit taken off there' (quoted in Cooper, 1975, 60). Thus, in the mid 1970s, differences in the level of service amongst regions in England were still conspicuous. In an analysis of expenditures in 1972–3, Rickard (1976) demonstrated that the highest expenditures were in the Thames Regions (Table 4.3) and that within the regional health authorities (RHA) the 'teaching' health districts spent more per head of population than non-teaching districts. (A teaching health district is one which includes a teaching hospital.) The variations identified are net of regional cross-boundary movements by patients and of the above average costs of teaching hospitals and regional specialties, so that 'the remaining differences indicated the true level of services' (Rickard, 1976, 229).

In response to growing awareness of these inequalities and, indeed, growing criticism of them, the Department of Health and Social Security established a Resource Allocation Working Party. It was set a difficult task. It was asked:

to review the arrangements for distributing NHS capital and revenue to regional health authorities, area health authorities and districts respectively with a view to establishing a method of securing as soon as practicable a pattern of distribution responsive objectively, equitably, and efficiently to relative needs and to make recommendations. (Gt Britain, 1976a, 5)

The underlying objective of these terms of reference was to create equal opportunity of access to health care for people at equal risk, to be achieved by selecting 'criteria which are broadly responsive to relative need, not supply or demand and to employ these criteria to establish and quantify in a relative way the differentials between geographical locations' (Gt Britain, 1976a, 7).

Table 4.3: *Per Capita* Expenditure of English Regional Health Authorities (1972–3)

Regional Health Authority	%[a]
Northern	+ 7.72
Yorkshire	+ 0.58
Trent	− 16.91
East Anglia	− 12.74
North West Thames	+ 12.76
North East Thames	+ 20.65
South East Thames	+ 10.54
South West Thames	+ 25.02
Wessex	− 8.56
Oxford	− 8.97
South-western	+ 1.50
West Midlands	− 16.90
Mersey	+ 15.30
North-western	− 2.74

Note:
a. % variation from the national average.
Source: Amended from Rickard (1976).

The revenue allocations to the regions can be taken to illustrate the RAWP approach. The determination of the level of revenue appropriate to each authority involves three steps:

1. For each authority a 'revenue target' is identified; this is the level of revenue monies an authority should receive according to its needs (Figure 4.6).
2. The position of each authority in relation to its 'revenue target' is determined; some will be above target (so-called over-funded authorities) and some below target (under-funded authorities).
3. The rate at which authorities can move from their existing positions to their 'revenue targets' is determined in the light of political and financial circumstances.

The key to the whole strategy is, therefore, the determination of the revenue target. The starting point is the mid-year population estimate and the seven sectors of NHS activity covered by the formula (Table 4.4). Significantly, general practitioner services are excluded, except for their administration costs. Each of the seven

Figure 4.6: The Build-up of a 'Revenue Target'

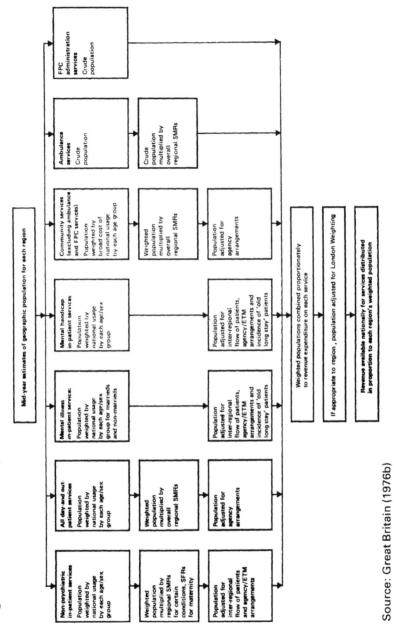

Source: Great Britain (1976b)

populations is then weighted using criteria of need defined by the Working Party. These are usually combinations of age, sex, mortality, fertility and marital status. The seven weighted populations are then combined in proportion to national revenue expenditure on each of the seven elements, and adjustments are made to take account of cross-boundary movements by patients. If appropriate, the population is adjusted to take account of extra costs in London, and finally the revenue available nationally is distributed in proportion to each regional authority's combined weighted population.

Table 4.4: NHS Expenditure by Programme Sector (Percentages)

Non-psychiatric inpatient	55.9
All day and outpatients	13.4
Community health	8.8
Ambulances	3.5
Mental illness inpatients	12.2
Mental handicap inpatients	5.7
Family Practitioner Committees Administration	0.5
	100.0

Source: Buxton and Klein (1978).

The effects of population weightings may be illustrated by looking in rather more detail at one sector – non-psychiatric inpatients services – which accounts for 56 per cent of NHS expenditure and thus represents the largest element in the overall resource allocation formula. RAWP recommended two sets of weights to be applied to the base population, one to reflect age and sex structure, the other the level of illness. The Standardised Mortality Ratio (SMR) was used as a surrogate for illness, because the Working Party could not find adequate data to measure morbidity directly. The effect of applying these weightings to the non-psychiatric inpatient population is demonstrated in Figure 4.7 which shows the impact of weightings on RHA populations. Oxford RHAs calculated notional population is 10 per cent less than its actual population while in contrast the North-Western RHAs notional population is 10 per cent greater than its actual population, implying that there is a greater need in the latter than the former.

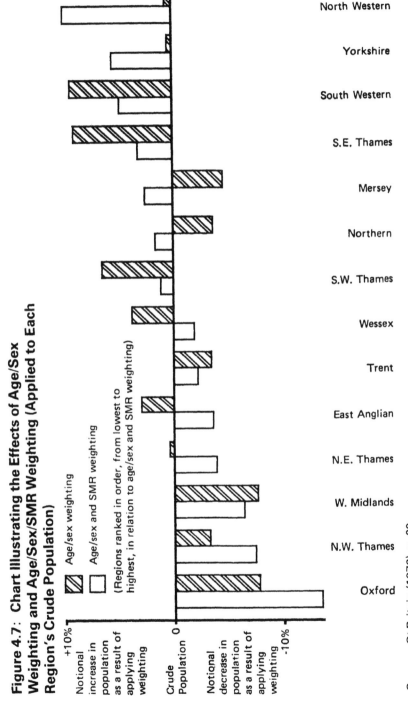

Figure 4.7: Chart Illustrating the Effects of Age/Sex Weighting and Age/Sex/SMR Weighting (Applied to Each Region's Crude Population)

Source: Gt Britain (1976), p. 20.

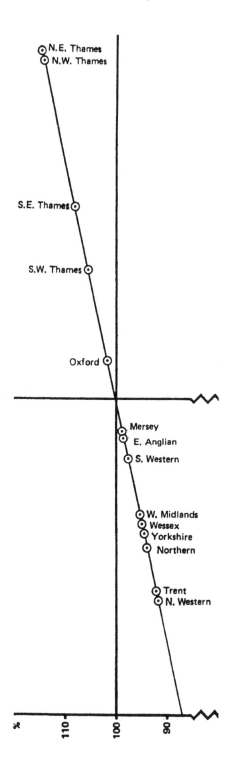

Figure 4.8: The Position of Each RHA in Relation to Its Revenue Target

Source: Gt Britain (1976b).

Similar adjustments are made in other elements (sectors) of the formula, which are then combined in proportion to the national revenue expenditure as listed in Table 4.4. Figure 4.8 shows the resultant position of each regional health authority in relation to the revenue target represented by the horizontal line. The position of each RHA indicates their distance from their target: those RHAs lying above the line are 'over-funded', and those below it 'under-funded'. The North-Western RHA is, therefore, the RHA worst provided, and the NETRHA the best provided. RAWP's intention was thus to allocate NHS revenue resources so that all authorities would ultimately lie along the horizontal line, that is all authorities would receive the revenue associated with their revenue target.

Whilst this discussion has concentrated on the distribution of revenue to RHAs the same process (although slightly modified) was to apply to the distribution of revenue by RHAs to their AHAs and Districts. It is these allocations which have had the biggest impact on the health services within cities and especially on the inner city. At a national level the practical impact of RAWP has been to divert funds away from the south-east to the north and west. Within regions the major impact has been to divert funds away from inner city areas to suburban and outer metropolitan authorities.

NETRHA provides a particularly clear example of the impact of the RAWP formula. Table 4.5 illustrates the formula's effects within this regional health authority. The pace of change from existing allocation to target allocations is a matter of political and administrative judgement, and most authorities have proceeded more slowly towards targets than was advocated in the original RAWP proposals. The principal reason why progress towards targets has been sluggish is the severity of the overall resource constraint in the NHS, which has limited the scope for reallocation using additional funds. Instead, authorities have been forced to consider changing existing allocations and this has proved to be difficult in practice.

Not surprisingly, then, the publication of the RAWP report created a storm, with those standing to lose most criticising it most (especially consultants and administrators from the London teaching hospitals), whilst those with resource gains gave a tentative welcome for the proposals (see Woods, 1982).

On a technical level and in the terms of RAWP itself, the question of the population weightings, and especially those applied to non-psychiatric inpatient services, can be raised. How adequate

Table 4.5: Illustrative Effect of RAWP Formula on Revenue Allocations in NETRHA (1978)

	Essex	B+H[b]	C+I[c]	C+EL[d]	E+H[e]	R+WF[f]
			Area Health Authority			
Revenue Target £000s[a]	121,284	34,672	50,218	66,067	34,197	52,917
Allocation £000s[a]	96,862	31,059	72,948	82,666	34,083	43,680
Allocation as % of target	79.8	89.5	145.2	125.1	99.6	82.4

Notes:
a. All prices are November 1977 levels.
b. B + H: Barking and Havering.
c. C + I: Camden and Islington.
d. C + EL: City and East London.
e. E + H: Enfield and Haringey.
f. R + WF: Redbridge and Waltham Forest.
Source: Appendix D, letter of 7 December 1978 to Area Administrators from Regional Treasurer, North East Thames RHA.

are age–sex utilisation and standardised mortality ratio weights as indicators of need? Some critics have no doubt that standardised mortality ratios, at least, are of little use because no consideration was given to the effects of social deprivation on health care needs (Avery Jones, 1976). Fox (1978) wanted to know if a mortality rate 10 per cent greater than average implied a 10 per cent greater burden on health services, thus implicitly questioning the unstated assumption of linear input–output relations. Geary (1977) demonstrated that in area health authorities standardised mortality ratios are subject to random fluctuations over time because they are based on small numbers of deaths, thus raising the issue of the actual measurability of local health status using these indicators. The problem, of course, is finding alternative measures of health status. Mortality data was used because the Working Party believed that morbidity data would mirror the pattern of mortality. Using data for the standard regions of England, Forster (1977) has questioned the significance of this relationship. A number of reviews has identified other technical deficiencies in the measurement of need (see, for instance, Buxton and Klein, 1978; Bevan *et al.*, 1980; Snaith, 1978), but there are other non-technical issues which have also been raised. First, the decision by RAWP to combine the seven sectors of NHS activity used in the formula in proportion to existing levels of national expenditure on them has been questioned (see Eyles *et al.*, 1982). It is by no means clear these proportions represent the desired or appropriate distribution of expenditures and, indeed, as we show in Chapter 6, at the time of RAWP's publication the DHSS was actively reviewing the distribution of expenditures by service sector. The Black Report (DHSS, 1980) highlighted this deficiency and recommended that proposed, rather than existing, sector expenditure distributions should be used in the formula.

A second major issue stems from the fact that RAWP says nothing about the pattern of services purchased by allocated financial resources; in its own words RAWP was, 'concerned with the means rather than the end. We have not regarded our remit as being concerned with how resources are deployed' (Gt Britain, 1976a). This, however, is a central issue for health authorities who find themselves 'overfunded', for forcing them towards a lower revenue target effectively restricts the scope they have to make changes in resource mixes which are historically determined and may be inappropriate in the contemporary context. Local medical hegemony is a major barrier to any attempts to enforce a redistribution in existing budgets. Inner city health authorities are the best example of this. With extremely high costs deriving from the historical concentration of teaching hospitals, coupled with declining local populations, they appear, according to the RAWP formula, as overfunded, but this obscures deficiencies in other aspects of local services like primary and community services on which inner city populations are heavily dependent. Attempts to redirect expenditures from existing resource allocations are problematic in the face of medical opinion which is not prepared to reduce spending on some services to finance expansion of others. Allocating health service resources is not, therefore, just a technical matter; technical problems of measurement are involved but, above all, resource allocation is a political issue. The adoption of 'need' as a distributive criterion, and its measurement using a complex formula, create the impression of a rational scientific process, but the specification of the formula and the methods of measurement adopted are, of course, related to the social, political and bureaucratic context in which they are formed. In Chapter 7 we return to this theme and expand our critique by placing the development of the policy in a wider context. In its own terms, however, RAWP represents an ambitious effort to achieve spatially equitable resource allocations and as such is an outstanding practical example of welfare geography.

Conclusions

In this chapter we have reviewed a number of studies which are examples of the economic perspective identified in Chapter 1. In this perspective the consumption of health services is regarded as beneficial, the major issue being the allocation and distribution of services. Implicit is the idea that the input of health care services produces health benefits. However, there is evidence that this is not so, that the increasing consumption of medical services does not necessarily produce a concomitant increase in health, and may even cause the individual and society harm (Illich, 1975). To discover if particular medical procedures have any beneficial impact Cochrane (1972) advocated the use of randomised controlled trials. In this way what Cochrane called the 'effectiveness' of medical therapy could be assessed by comparing the outcome of patients randomly allocated to groups, one of which received the specific therapy. Cochrane (1972) also advocated studies to investigate the efficiency of clinical practice, to guide our judgements about, for instance, hospital lengths of stay and in wider terms the balance between hospital based and community based care. The need for such studies stems from the fact that current practice owes more to historical factors than to precise knowledge about appropriate lengths of stay or the balance of care. More is known about the relationship between resource inputs and health service activity – as measured, perhaps, by inpatient and outpatient consultations – but it is more problematic to identify any specific health outputs or outcomes as measured by, say, reduced morbidity or mortality (see Fenton Lewis and Modle, 1982, for a discussion of the distinction between inputs, activity, output and outcome). In a study of the relationship between age specific mortality rates, health facilities, environmental and dietary factors in countries with a GNP *per capita* of at least $2000 in 1972, Cochrane *et al.* (1978) found that countries with high inputs of health care were not associated with low levels of mortality. On the contrary, those countries with the highest prevalence of doctors had the worst mortality rates in younger age groups. As the authors comment it is doubtful this is a

Figure 4.9: *Per Capita* **Health Expenditure and Perinatal Mortality in Countries of the OECD**

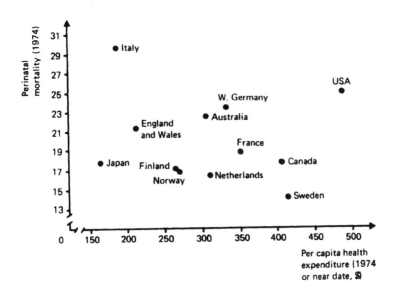

Source: Gt Britain (1979b), p.25.

Figure 4.10: Doctors Per 10,000 Population and Perinatal Mortality in Countries of the OECD

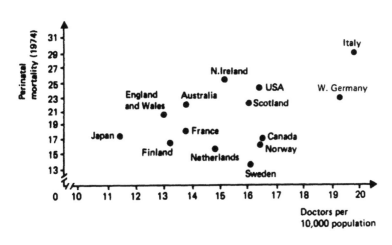

Source: Gt Britain (1979b), p. 25.

causal relationship, but it does indicate that there is no simple positive relationship between health care input and health output. The same conclusion was reached by the Royal Commission on the NHS (Gt Britain, 1979b) which examined the relationship between *per capita* health expenditure and doctor:patient ratios and perinatal mortality in the countries of the OECD (Figures 4.9 and 4.10). They found the country with the highest spending (the USA) had the worst perinatal mortality rate with the exception of Italy. In the case of doctor:patient ratios, Figure 4.10 shows again that higher ratios are associated with high perinatal mortality rates. Again it would be dangerous to attach causal significance to these results since mortality measures may be inappropriate measures of outcome, and *per capita* expenditure tells us nothing about the nature of services purchased nor about the distribution of those expenditures amongst the population.

The general point, however, is that as increasing resources are devoted to health care, and as attempts are made to optimise the distribution of services purchased, this process continues with only partial knowledge of the relationship between inputs and health outcomes. As Eyles *et al.* (1982, 241) observe:

> the calculation of input–output coefficients implies some precise knowledge of the relevant 'production process', which may be fairly straightforward in the case of how much coal, iron ore and limestone is needed for one ton of pig iron, but less so with respect to the most efficient way of ensuring that a newly born child survives to its first birthday.

In the absence of precise input–output coefficients the combination of health care inputs which constitutes our health care system is derived from historical practice and is also a reflection of prevailing concepts of health and illness. The nature of the health care system itself also becomes a major barrier to change because it has its own set of ethics, practices and vested interests, which in turn are rooted in the nature of society itself. It is these issues which form the focus of our next two chapters.

5 PRACTICAL RESPONSES: THE EFFECTS OF CONSTRAINTS

Introduction

Throughout our preceding discussion of approaches to facility location and service utilisation we continually hinted at constraints in the real world which necessarily deflect from optimal location patterns or solutions. The power of these constraints ensures that locational optima are the exception rather than the rule, and in this chapter we explore how three types of constraints influence the practical responses of health care providers and users. The three types of constraint which we distinguish are: historical; social; and professional. The first of these is concerned with the problem of adapting existing health care systems to changing demands. It is a consideration of the impact historical inertia in health care systems has upon the planning of patient services. The emphasis is upon how the inherited size and location of health care facilities dictates, in part, how future systems will develop. In contrast our second category is concerned with how social characteristics such as race, income, religion and class affect the individual's health care opportunities in relation to his health experiences. Our final set of constraints are in a sense a link between the other two, for in the case of professional constraints we are concerned with how the objectives and values of health care professionals influence the shape of health care systems.

To an extent these distinctions are arbitrary since the constraints usually interact in particular places to produce particular health care systems. Further, there are cultural and systemic forces at work that affect the operation of these constraints (see Chapter 6). Even so, by adopting this classification it is easier to identify the role played by each set of constraints in the creation of complex health care delivery systems which, in part, can be viewed as particular manifestations of their interaction.

Historical Constraints: The Example of Acute Hospital Services in London

Health care planners are normally confronted with a set of pre-

145

Figure 5.1: District Health Authorities in Greater London, After 1 April 1982

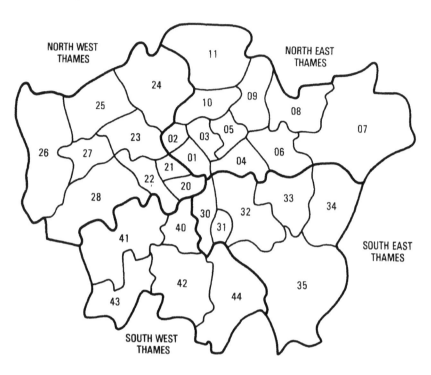

0   16 km *(approximate scale)*

▬▬▬   Regional Health Authority Boundary

────   District Health Authority Boundary

DISTRICT HEALTH AUTHORITIES

North East Thames
01 Bloomsbury
02 Hampstead
03 Islington
04 Tower Hamlets
05 City Hackney
06 Newham
07 Barking, Havering
   and Brentwood
08 Redbridge
09 Waltham Forest
10 Haringey
11 Enfield

North West Thames
20 Victoria
21 Paddington and North Kensington
22 Hammersmith and Fulham
23 Brent
24 Barnet
25 Harrow
26 Hillingdon
27 Ealing
28 Hounslow and Spelthorne

South East Thames
30 West Lambeth
31 Camberwell
32 Lewisham and North Southwark
33 Greenwich
34 Bexley
35 Bromley

South West Thames
40 Wandsworth
41 Richmond, Twickenham
     and Roehampton
42 Merton and Sutton
43 Kingston and Esher
44 Croydon

existing facilities, located perhaps for reasons which are no longer relevant, linked together by informal as well as formal professional and administrative arrangements. We have already referred to the inertia in general practitioner locations identified by Phillips (1981) and Knox (1978; 1979), but nowhere is this better demonstrated than in the case of London's acute hospital services which are under tremendous pressure to change from a pattern that is the product of hundreds of years of development.

The pressure for change in these services stems from the post-war migration of inner London's population to the suburbs and the outer-metropolitan counties beyond the green belt. Populations are able to relocate much more rapidly than hospital systems can, and health authorities in London have been faced with the need to 'catch up' with the population which has moved out. At the same time they are also faced with the reciprocal question of what to do with the hospitals which remain in inner London bereft of the populations which spawned them. In 1979 a detailed analysis of this problem was published by a specially created body called the London Health Planning Consortium (LHPC). The following year the LHPC produced a second document which carried the analysis further and, in conjunction with the University of London, made specific illustrative recommendations for the location of acute services in hospitals which would (it was hoped) protect the needs of medical education (which is the University's responsibility) provided through medical colleges linked to NHS hospitals. The LHPC is a non-executive, though extremely influential, body composed of representatives from the four Regional Health Authorities which cover the London region (the Thames Regions as they are called hereafter, Figure 5.1), the Department of Health and Social Security, the University Grants Committee (UGC) and the Postgraduate Hospitals (which have a special role in the training of hospital consultants). The reports produced by the Consortium provide an excellent example of the effects of historical constraints, for whilst the task confronted by the LHPC can be characterised as one of pouring new wine into old bottles, the volume of wine and the number of bottles have also changed.

London's present day hospital system is essentially a product of the nineteenth century (see Abel-Smith, 1964; Clarke-Kennedy, 1979). Whilst the population remained concentrated in the immediate vicinity of inner London's hospitals there was no great locational imbalance, but when in the present century the population began to move away from central London as a consequence of the war, and especially post war planning, the inner London hospitals became increasingly isolated from the populations they served. Cowan (1967) drew attention to the implication this changing population distribution had for the geographical disposition of hospitals, by comparing population density and hospital bed densities in 1901 and 1965 (Figure 5.2). He concluded, 'that London's hospital provision lags some sixty-five years behind the growth and spread of population' (Cowan, 1967, 112–13).

Cowan's model has its limitations. For example it takes no account of regional specialties which have high population thresholds and, therefore, serve larger geographical areas, but the undoubted imbalance which he highlighted is currently reflected in the acute hospital bed:population ratios (Table 5.1) and the movement of patients from the NHS administrative area where they live to other areas where they receive treatment. A comparison of Table 5.1 with Figure 5.1 reveals that it is the inner city area health authorities like Camden and Islington and the City and East London which have the highest bed:population ratios, and it should be noted these are ratios which specifically exclude beds allocated to regional specialties. In response to these differences in provision patients move across formal administrative boundaries adding significantly to the workload. For example in the City and East London AHA (T) in 1974, up to 25 per cent (net) of all people admitted to beds in local acute specialties were from outside that AHA (City and East London AHA (T), 1978). Studies of outpatient utilisation have shown even larger proportions flowing into central London from the suburbs. In the Tower Hamlets Health District it was found that of all new outpatient attenders in a six month period, only 33 per cent were locally resident (Woods, 1979). (This is not a net figure; some Tower Hamlets residents would be treated in other health districts and thus partially compensate the influx.) There was also considerable specialty

Figure 5.2: London Population Density 1901 and 1961 (Persons Per Sq. Mile) and Bed Density 1965 (Beds Per Sq. Mile) by Distance from Centre

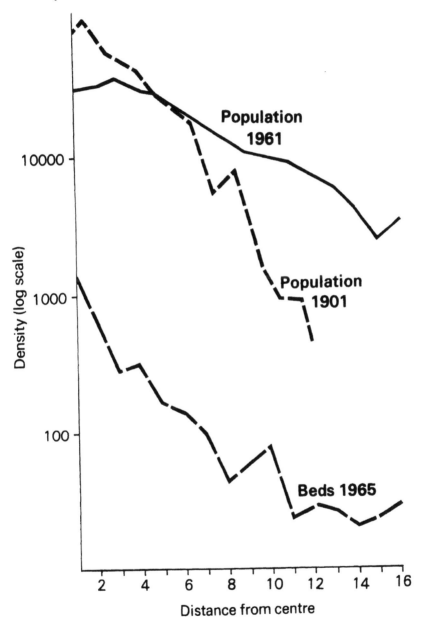

Source: Cowan (1967).

variation, with as expected the regional specialties having the smallest proportions of their workload drawn from local residents. Similar results have been recorded in other special surveys in central London hospitals (St Thomas' Health District, 1978; Department of Community Medicine, Middlesex Hospital, Medical School, 1978).

Table 5.1: Acute Beds (Excluding Regional Specialties) Per 1,000 Residents in Each Area Health Authority Within the North East Thames Regional Health Authority, 1975

North East Thames	3.24
City and East London AHA (T)	5.12
Camden and Islington AHA (T)	4.52
Enfield and Haringey AHA	3.46
Redbridge and Waltham Forest AHA	2.95
Barking and Havering AHA	2.75
Essex AHA	2.11

Source: North East Thames RHA Strategic Plan, (revised draft) 1976.

It is in this context that the LHPC have set out to refine Cowan's basic model and determine 'how the present imbalance would need to be adjusted to bring about a more reasonable distribution of acute hospital services consistent with the needs of the population' by 1988 (Gt Britain, 1979a, 28). Two basic questions have, therefore, been addressed:

1. How many beds in each clinical specialty are required in each health district?
2. Which hospitals should those beds be in to ensure an equitable distribution of local services, a convenient distribution of regional services and a service that is suitable for the purposes of medical education (which requires access to patients in all specialties and in numbers commensurate with the number of students).

The first of these questions was answered by undertaking a quantitative analysis (Gt Britain, 1979a; see Figures 5.3 and 5.4) and the second by subjective judgements, which not surprisingly have proved controversial. Three steps can be identified in this process. First, the determination of 'notional bed values'; that is the

Figure 5.3:   Calculation of Notional Bed Values

```
┌─────────────────────┐         ┌─────────────────────┐
│ 1988 projected      │         │ 1988 population     │
│ age/sex specific    │         │ projections for     │
│ hospitalisation     │         │ each area of        │
│ rates for each      │         │ residence (by       │
│ specialty in each   │         │ age/sex group)      │
│ area of residence   │         │                     │
└─────────────────────┘         └─────────────────────┘
```

| Projected 1988 turnover interval for each specialty | Projected 1988 age/sex specific lengths of stay for each specialty | Projected 1988 cases (for each age/sex group in each specialty) for each area of residence |

```
┌─────────────────────┐
│ Beds required       │
│ to treat projected  │
│ cases in each       │
│ specialty for       │
│ each area of        │
│ residence           │
│ 'Notional Bed Values' │
└─────────────────────┘
```

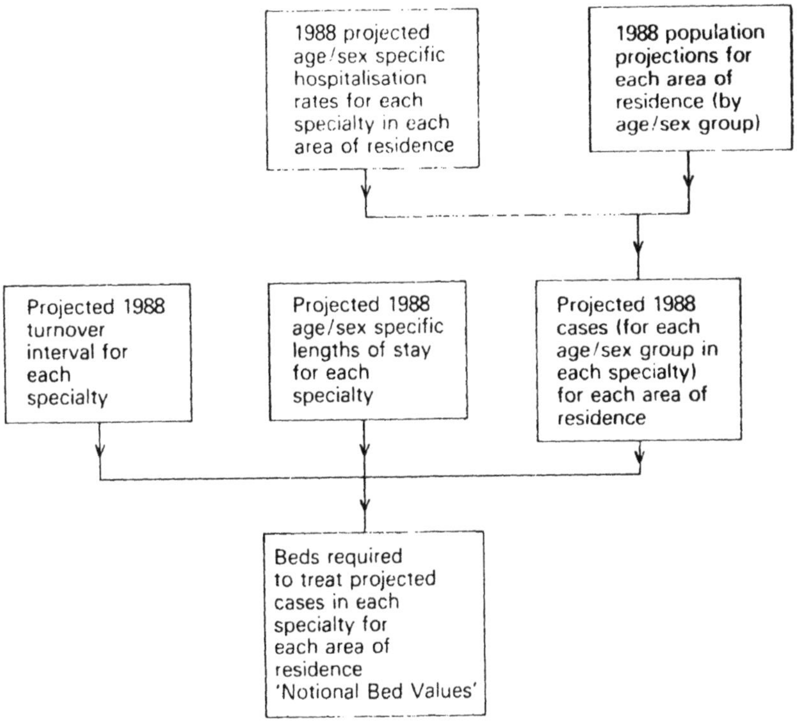

Source: Gt Britain (1979a).

number of beds required by each specialty in 1988 to treat the caseload associated with the population in each of the areas of residence (health district), is derived (Figure 5.3). Secondly, the notional bed values are allocated amongst the health districts on the basis of where people receive treatment rather than where they live (Figure 5.4). Thirdly, decisions as to which hospitals should house the new specialty allocation are made.

The starting point for the derivation of notional bed values was an estimate of the age and sex structure of the population in 1988. Since hospitalisation rates for each specialty can be calculated on the basis of past utilisation patterns it is possible to project the caseload expected from a future population with known age-sex characteristics. By combining this caseload with the expected length of stay of patients in 1988 (again based on projections of past trends) and the patient turnover interval (length of time between

Figure 5.4:   Allocation of Notional Bed Values to Health Districts

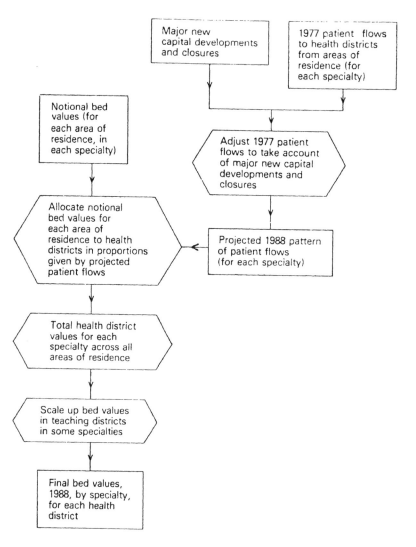

Source: Gt Britain (1979a).

patients) the number of beds required by each specialty can be calculated. Some modification is made to take account of the greater needs of certain inner city health districts where adverse socioenvironmental factors are said to increase the need for hospitalisation. But because patients are not always treated in the health district where they live the distribution of notional bed values is adjusted to take account of the expected cross boundary movement of patients in 1988. New hospital developments and closures are taken into account, and additional beds for teaching purposes are allocated to health districts containing teaching hospitals.

The crucial element in this process is the determination of the notional bed value, and as the LHPC itself recognised, their analysis was entirely dependent upon the accuracy and validity of their estimates and projections. Will the 1988 age-sex structure and distribution match that predicted by the LHPC in 1979? Are the projected lengths of stay, patient turnover intervals, etc. likely to be achieved? Are modifications to take account of greater need in some inner city health districts sufficient to meet those needs? (Woods, 1982). Should any of the answers to these questions prove to be different from the LHPC's assumptions then the analysis will be correspondingly inaccurate. This is particularly problematic for inner city health districts, for these are calculated to suffer the largest reduction in the number of hospital beds (see Figure 5.5). Overall, the number of beds required in 1988 is predicted to be 4,934 fewer than in 1977 (11 per cent of the 1977 level), but some inner city teaching health districts are predicted to require 20 per cent fewer beds. Most of the bed reductions in such authorities will be in the local acute specialties. Table 5.2 presents the number of beds predicted to be needed in each specialty in the Tower Hamlets Health District. Largest reductions are seen in General Medicine, General Surgery and Paediatrics, whilst the regional specialties like Neurology, Cardiology and Neurosurgery would receive additional beds. It is this change in the distribution of beds by specialty which has the biggest implications for the location of beds in particular hospitals and for medical education. The question of medical education is inextricably involved in this proposed reorganisation; medical students require access to both the less common conditions treated in specialist regional units and to large numbers of hospital patients with more common conditions if they are to obtain adequate clinical experience. Any decision to change the distribution of hospital beds has, therefore, considerable implications for medical education. Given that the medical schools and colleges in London are generally attached to inner city teaching hospitals this means that these hospitals must remain at the heart of any revised system, though they are still likely to be affected as a result of the joint deliberations between the University and the LHPC. These joint deliberations show not only how the past complicates the present, but also how the process of decision making is influenced by the power of the medical profession to mould policy (see below).

Table 5.2: Projected Bed Requirements in 1988 and Allocation in 1977 in Tower Hamlets by Specialty

	1977	1988	% change
General Medicine	320	193	− 39.6
General Surgery	262	182	− 30.5
Paediatrics	46	21	− 54.3
Trauma and Orthopaedic	113	142	+ 25.6
Gynaecology	57	53	− 7.0
Chest, Infectious, Rehab/Rheumatology, Dermatology and STD	66	51	− 22.7
ENT, Opthalmology Dental Surgery, Urology	108	105	− 2.7
Regional Specialties (Neurology, Cardiology, Thoracic Surgery, Neurosurgery, Plastic Surgery)	126	158	+ 25.39
Total	1098	905	− 17.5

Source: DHSS (1980).

The primary recommendations of the LHPC and Flowers Report (University of London, 1980) was that six medical schools should be formed from the amalgamation of 34 existing teaching institutions. These were to be linked to designated 'University Hospitals' and would provide the clinical training for medical students. Some would be Postgraduate Specialist Hospitals, some the traditional teaching hospitals where regional specialties would be concentrated and the remaining hospitals located in the suburbs designated as 'complementary hospitals' ensuring that medical students had access to large numbers of patients in the local acute specialties (Table 5.3).

Figure 5.5: The London Health Planning Consortium's Proposed Percentage Changes in the Distribution of Acute Hospital Beds (All Specialties) by Health District in the Thames Regional Health Authorities

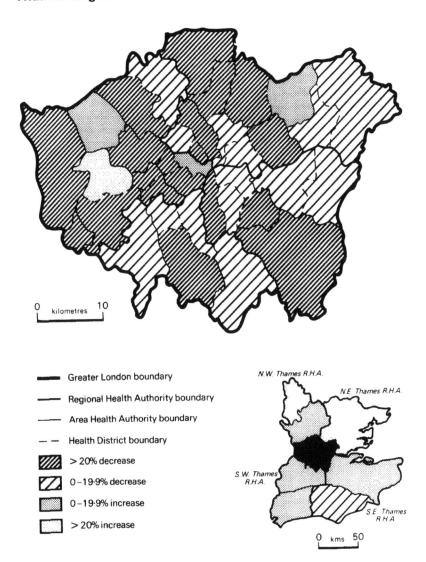

Greater London boundary

Regional Health Authority boundary

Area Health Authority boundary

Health District boundary

> 20% decrease

0–19·9% decrease

0–19·9% increase

> 20% increase

Source: based on Table 2, DHSS, 1980, 6.

Table 5.3:   Medical Schools and University Hospitals Proposed by the Flowers Report, 1980

Medical School	*University Hospitals*
1. *St George's School of Medicine and Dentistry* Formed from: St George's Hospital Medical School Royal London School of Dental Surgery Institute of Cancer Research	St George's Group of Hospitals Atkinson Morley Hospital St James's Hospital St Helier Hospital Queen Mary Hospital, Roehampton Royal Dental Hospital Royal Marsden Hospital
2. *St Mary's and Royal Postgraduate Joint School of Medicine and Dentistry* Formed from: St Mary's Hospital Medical School Royal Postgraduate Medical School Institute of Dental Surgery	St Mary's Hospital Central Middlesex Hospital Hammersmith Hospital Ealing Hospital Eastman Dental Hospital
3. *Charing Cross School of Medicine* Formed from: Charing Cross Hospital Medical School Cardiothoracic Institute	Charing Cross Hospital National Heart and Chest Hospitals St Stephen's Hospital

4. *The Harvey School of Medicine and Dentistry*
Formed from:
St Bartholomew's Hospital
    Medical College
The London Hospital Medical College
Institute of Opthalmology
	Queen Elizabeth Hospital
	    for Sick Children
	St Bartholomew's Hospital
	The London Hospital, Whitechapel
	    and Mile End
	Homerton Hospital
	Newham Hospital
	Moorfields Eye Hospital

5. *University College School of Medicine and Dentistry*
Formed from:
Middlesex Hospital Medical School
Royal Free Hospital School
    of Medicine
UCL Faculty of Clinical Sciences
London School of Hygiene and
    Tropical Medicine
Institute of Neurology
Institute of Child Health
	University College Hospital
	Middlesex Hospital
	Royal Free Hospital
	Whittington Hospital
	Royal National Orthopaedic
	    Hospital, Stanmore
	National Hospital for
	    Nervous Diseases
	Hospital for Sick Children,
	    Gt Ormond St

6. *The Lister and St Thomas' Joint School of Medicine and Dentistry* Formed from: King's College Hospital Medical School Guy's Hospital Medical School Institute of Psychiatry	Guy's Hospital King's College Hospital Lewisham Hospital St Thomas' Hospital Maudsley Hospital Bethlem Royal Hospital

Source: University of London (1980), pp. 48–9.

In this way the Flowers report believed that students would have access to the necessary variety of clinical experience and at the same time these arrangements would dovetail with the objectives of the

LHPC. Inevitably the proposed arrangements meant some medical schools would either close or lose their separate identities. It was chiefly for this reason that agreement on the Flowers' proposals proved illusory as the individual institutions made representations to the University in order to protect their interests. The Joint Medical Advisory Committee (JMAC), which represents the views of the medical schools through their Deans, and the Joint Planning Committee (JPC) of the University and of the Senate were the two bodies through which these representations were made. The JMAC advocated a slower, more evolutionary approach than that proposed by Flowers'; they agreed, however, that some mergers were necessary, and the JPC took the JMAC suggestions on Flowers' as the basis of its own recommendations to the Senate of the University.

This detailed discussion led to another working party under Professor Le Quesne which confirmed many of Flowers' findings on costs. In the light of the Le Quesne Working Party's data the JPC subsequently recommended to the Senate that consortia be created (Table 5.4) to enable collaboration and mergers in those places where consortium members could reach agreement. The Senate however, was not prepared to accept this scheme. It only agreed to recognise the geographical groupings as indicative and subject to modification. Following further discussions a definitive distribution of medical students was proposed to the Senate in December 1981. The Senate resolved to accept these and thus they became the University's policy on the future of medical education (Table 5.4). It remains unclear which of the existing stock of hospitals will be used by each of the new groupings, but as a general rule it seems larger hospitals will be retained in the belief that economies of scale derive from concentrating services on their sites, and there is no doubt historical locations will still form the basis of provision for London's acute hospital services. The political strength of the profession and the inertia of the existing system have clearly acted as powerful constraints on the implementation of change.

Social Constraints

Access to and use of health services is not only influenced by the relative locations of facilities and potential patients, but is also influenced by social characteristics such as age, sex, marital status, religion, ethnicity, class and income. Of particular relevance to this discussion is the evidence which shows how these various characteristics constrain accessibility opportunities, reflected most often in levels of service utilisation lower than would be expected. In some cases the apparent under-utilisation of services can be attributed to discrimination against particular individuals (e.g. on the basis of race). For instance, Cordle and Tyroler (1974) have drawn attention to differences in hospital utilisation rates by race in Charleston County, South Carolina. In their study of hospital admissions in 1963 they found that the admission rate for whites was 4.5 times that for blacks. Those blacks who were admitted to hospital tended to be more seriously ill than whites, the major factor in explaining an in-hospital mortality rate for blacks eight times that experienced by whites.

The Chicago Regional Hospital Study (CRHS) also highlighted the effects of race and income on access to medical care. In the United States the distribution of general practitioners (or physicians as they are usually known) is closely correlated with race and wealth. Wealthy, white neighbourhoods tend to be doctor-rich and low income black neighbourhoods doctor-poor areas (Shannon and Dever, 1974). De Visé (1973) reported that of the general practitioners in Chicago's black poor neighbourhoods few could admit patients to private hospitals because they lacked staff privileges which give the right to admit. As a result, black patients became dependent upon the one major public medical service, Cook County Hospital, because it was the one hospital unable to deny admission.

De Visé (1973) has summarised the impact of race and other social characteristics on access to health care in the United States as follows:

> If you are either very poor, blind, disabled, over 65, male, white or live in a middle or upper-class neighbourhood in a large urban centre, you belong to a privileged class of health care recipients and your chances of survival are good. But, if you are none of these, if you are only average poor, under 65, female, black or live in a low income neighbourhood, small town, or rural area, you are a disenfranchised citizen as far as health care rights go, and your chances of survival are not good. (De Visé, 1973, 1)

In the United Kingdom the creation of the National Health Service was intended to eliminate inequalities in access and service

Table 5.4: Future Medical Schools in London?

(i)   *Consortium Proposed by the Joint Planning Committee of the University of London*

West London
  Charing Cross HMS and Westminster MS
  St George's HMS

South London
  Guy's HMS and St Thomas' HMS
  King's College and King's College HMC

East London
  St Bartholomew's HMC and The London HMC

North Central London
  University College, Middlesex HMS, St Mary's HMS
  Royal Free HSM

Source: Paragraph 56, Minute 747 of the Senate of the University of London, 25 March 1981.

(ii)   *Proposed Definitive Distribution of Medical Students*

	Preclinical students	Clinical students
West London		
Charing Cross HMS and Westminster MS	155	175
St George's HMS	150	160
South London		
Guy's HMS and St Thomas' HMS	195	225
King's College and King's College HMS	105	115
East London		
St Bartholomew's HMC and The London HMC	200	250
North Central London		
Royal Free HSM	100	100
St Mary's HMS	100	110
The Middlesex HMS and University College	195	215
	1200	1350

Source: Paragraph 43, Minute 483 of the Senate of the University of London, 16 December 1981, and Appendix SPC1 Annex A.

use which stemmed from social characteristics. Even so, there is evidence that various social constraints still operate, perhaps not as apparently as in the USA, but just as effectively. For instance, Butler and Morgan (1977) have discussed the impact of marital status on hospital use, showing that single people are admitted to and stay in hospital more frequently and longer than married persons. Two alternative hypotheses were advanced in explanation: first, the incidence of illness is differentially distributed between marital status groups; and secondly, that clinicians are more likely to admit single people and keep them in hospital longer if they believe such patients lack family support at home. But perhaps the most significant social constraint which still affects access and use of service is social class. In Chapter 3 we described class inequalities in health as measured by mortality and morbidity. Such inequalities are also found in respect to health service utilisation. Although Rein (1969) asserted that the NHS had in fact overcome inequalities in use by different social classes, Titmuss (1968) argued that the middle class still received better medical care because it knew how to get the most out of the NHS. Since then the debate has continued, the major problem confronting research in this field being to relate data on need with that on use. In principle, high levels of need should be paralleled by high levels of use, but in practice it has proved difficult to acquire data to test if this obtains. Data on GP consultation rates are available from the General Household Survey (GHS) and have been fully discussed in the Black Report (DHSS, 1980). Table 5.5 presents the data from the Black Report which shows higher consultation rates for manual workers (for both men and women) than non-manual workers when all age groups are considered together. But this, of course, does not mean that manual workers are using GPs at the rate which the mortality and morbidity data suggest they should. Various attempts have been made to resolve this question by deriving 'use:need' ratios for each social group (e.g. Brotherston, 1976; Blaxter, 1976). The Black Report followed the approach used by Brotherston (1976) and divided the number of GP consultations by the number of restricted activity days over a two week period for each social class as recorded by the GHS between 1974 and 1976 (Table 5.6). Quite clearly the ratios decline from social class I to social class V, suggesting that manual workers use general practitioner services less in relation to these needs than do non-manual workers and thus contradicting the crude utilisation rate.

More recently Collins and Klein (1980) have suggested that these results are a misinterpretation of the GHS data because they are based on inferences derived from two different populations, one reporting morbidity and the other use. Collins and Klein, therefore, reanalysed the original 1974 GHS data identifying four groups of

Table 5.5: Doctor Consultations: Persons Consulting a GP in a Two Week Reference Period by Sex and Socioeconomic Group (Rates per 1,000; GB, 1977; All Ages)

	Males	Females
Professional	69	120
Employees/managers	102	116
Intermediate	104	127
Skilled manual	103	134
Semi-skilled manual	112	143
Unskilled manual	138	157

Source: DHSS (1980), p. 96.

Table 5.6: Use:Need Ratios by Sex and Socioeconomic Group

	Males	Females
Professional	0.23	0.23
Employees/managers	0.24	0.24
Intermediate	0.20	0.22
Skilled manual	0.18	0.22
Semi-skilled manual	0.20	0.20
Unskilled	0.17	0.19
All	0.19	0.22

Source: DHSS (1980), p. 97.

users of primary care services: those who reported no morbidity; those who reported acute illness; those who reported chronic sickness without restricted activity; and those also reporting chronic sickness causing restricted activity. For each of these health status groups they calculated the percentage of primary care users in each socioeconomic group. In order to check for the effect of age differences which could influence the rate of use, they also calculated a 'predicted use' rate which is the rate of use expected given the age composition of the groups. Table 5.7 presents a summary of the results on the basis of which 'we can confidently conclude that Britain's primary health care system does not speak with an upper class accent' (Collins and Klein, 1980, 114). The basis for this statement, which contradicts the other evidence cited, is the absence of a relationship between class and 'actual' use in which the professional and managerial socioeconomic groups have higher utilisation rates than the manual socioeconomic groups. This is most clearly seen in the case of the 'actual' use made by males. In all health status groups except those 'not sick' the actual rate of use by the unskilled manual group is considerably higher than socioeconomic group 1. The significance of this evidence is not completely clear and in any case it represents only one aspect of service use. Consultation content is another important aspect – one which we, in fact, identified in Chapter 1 – and it has been suggested that middle-class patients obtain more from a GP consultation than their working-class equivalents. Cartwright and O'Brien (1976), for example, have shown that middle-class patients experience longer consultations and discuss more problems than do working-class patients.

In the case of hospital services there is little evidence with which to evaluate the effect of social class on utilisation. The GHS provides some data on outpatient attendances (see Table 5.8), but there is no systematic pattern of higher utilisation by the lower social classes. The only inpatient data of any value in this context comes from Scotland, where the social class of each inpatient is routinely recorded. Carstairs and Patterson (1966) found that hospital admission rates and length of hospital stay increased from social class I to social class V, and more recent Scottish data (Table 5.9) confirms this pattern. It is, however, with respect to preventive health care that the most convincing evidence of under-utilisation by manual workers and their families is found. Brotherston (1976) has shown late ante-natal booking to be more common in social classes III, IV and V than in social classes I and II. In 1975, 27 per cent of social class I women made an ante-natal booking after more than 20 weeks gestation, whilst the corresponding figure for social class V women was 40.5 per cent. The Second Report of National Child Development Study based on a 1958 cohort of children found a clear social class gradient in the uptake of immunisation against smallpox, polio and diptheria, as well as visits to see the dentist (see DHSS, 1980 and Table 5.10). Only 6 per cent of children aged seven with social class I father's had not been immunised against smallpox compared with 33 per cent of children with father's in social class V. The corresponding figures for non-attendance at the dentist were 16 per cent and 31 per cent.

In summary, there is, then, contradictory evidence that patients from social classes III, IV and V use general practitioners less than

Table 5.7: Users of Primary Care Services (Actual and Predicted) by Health Status and Socioeconomic Group

SEG^a	Males								Females							
	Not sick		Acute sick		Chronic sick (unrestricted)		Chronic sick (restricted)		Not sick		Acute sick		Chronic sick (unrestricted)		Chronic sick (restricted)	
	A^b	p^b	A	P	A	P	A	P	A	P	A	P	A	P	A	P
1	7.0^c	5.2	40.9	49.8	7.1	13.2	16.7	24.4	7.5	7.0	57.7	57.1	22.2	18.3	26.3	22.8
2	4.9	5.2	41.7	50.0	13.5	13.5	24.6	24.0	8.5	6.9	38.7	53.1	14.1	16.3	24.2	23.8
3	5.0	5.1	43.9	49.3	10.1	13.4	21.5	24.2	8.4	8.7	48.7	49.6	14.8	17.0	23.3	25.4
4	4.8	5.1	50.2	49.9	14.5	13.3	24.1	24.1	7.4	7.0	57.3	54.3	18.9	18.0	25.9	23.4
5	5.3	5.1	57.7	50.4	15.1	13.1	30.1	24.1	7.3	8.0	54.7	50.5	17.4	16.6	25.4	24.8
6	5.4	5.3	71.0	47.8	14.6	13.0	18.2	24.1	8.5	7.7	47.2	45.2	20.0	16.9	22.5	24.6

Notes:
a. SEG = Socioeconomic Group: 1 = Professional; 2 = Employees and Managers; 3 = Intermediate and Junior Non-manual; 4 = Skilled manual; 5 = Semi-skilled manual; 6 = Unskilled manual.
b. A = Actual use; P = Predicted use on basis of age distribution of GHS respondents.
c. All figures are the percentage users of primary care in each socioeconomic group, for each health status category.

Source: Based on Collins and Klein (1980), Tables III, V, VII, IX, pp. 113 and 114.

Table 5.8: Persons by Age, Sex and Socioeconomic Group Attending Outpatients in a Three Month Reference Period (Rates Per 1,000; Great Britain, 1974–7 data combined)

Socioeconomic group	Males				Females			
	0–14	15–44	45–64	65+	0–14	15–44	45–64	65+
1	96.2	79.0	87.9	132.2	76.9	103.3	101.6	104.2
2	96.3	94.4	98.4	112.0	67.4	96.9	110.3	97.4
3	113.0	100.7	122.3	142.0	77.9	107.3	123.8	134.5
4	89.5	112.9	117.0	123.5	75.1	96.4	112.6	122.7
5	82.5	102.8	132.0	122.5	68.6	86.6	114.1	121.9
6	108.2	133.5	113.7	104.7	58.9	90.9	107.6	130.9
All groups	94.2	105.4	115.8	123.2	72.6	98.0	114.2	123.7

Source: DHSS (1980), p. 102.

Table 5.9: Hospital Standardised Discharge Ratios (SDR) and Standardised Bed-Day Ratios (SBDR) by Social Class, Scotland 1971

Class	SDR		SBDR	
	Males	Females	Males	Females
I	79.5	95.9	63.7	92.5
II	80.9	98.0	73.3	93.6
III	94.0	90.4	93.9	91.0
IV	115.1	107.4	116.4	106.7
V	141.4	161.1	151.7	153.9

Source: DHSS (1980), p. 104.

Table 5.10: Use of Health Services by Children Under Seven Years of Age by Occupational Class of Father

	I	II	IIIm	IIIm	IV	V
Per cent never visited dentist	16	20	19	24	27	31
Per cent not immunised against						
Smallpox	6	14	16	25	29	33
Polio	1	3	3	9	6	10
Diptheria	1	3	3	6	8	11

Source: DHSS (1980), p. 106.

their needs suggest that they should. It is more clear that when they do use them, they tend to get less from their consultation than their middle-class counterparts. Hospital utilisation data is incomplete but suggests higher use of inpatient services by manual workers, though this pattern is not replicated in the case of outpatients, and the question of whether the gradient of use is steep enough to mirror the gradient of need remains open (Brotherston 1976). However, the clearest evidence that the non-manual worker uses the NHS more effectively than the manual worker is the differential uptake of preventive health care. The crucial question for the NHS is why do class groups behave differently? The answer lies in the perception of the cost and benefit derived from using a service, utilisation resulting usually when the perceived benefits exceed costs, perceptions which are, obviously, class related. Whilst the NHS does not make direct charges for most services, there are indirect costs in using services such as time, especially time off work which middle-class groups might find easier to absorb. Awareness of entitlements, knowledge of service availability and the ability to articulate needs and make demands all conspire in this process to produce differential utilisation. At the same time the NHS and its staff, especially its medical staff, are often organised and structured in accordance with middle-class values and assumptions which create barriers to access for people who 'are less able to express themselves in acceptable terms and who suffer from a lack of command over resources both of time and money' (DHSS, 1980, 117).

In addition the interpretation of illness and symptoms is a second critical influence. Illness behaviour is a central element in the sociology of medicine and health. We do not intend to review this broad field which:

> involves the study of attentiveness to pain and symptomology, examination of the processes affecting the way pain and symptoms are defined, accorded significance and socially labelled, and consideration of the extent to which help is sought, change in life regimen affected, and claims made on others. (Mechanic, 1978, 249)

We wish only to point out that illness behaviour depends not only on social structure and availability of care but also on the interpretation of illness by the sufferer. This interpretation is shaped by individual perception, personality and the immediate social environment.

Two groups of interest in this context are immigrants and women. Coombe (1976) indicates that some immigrants hesitate in seeking ante-natal care, and though there is a paucity of data relating use and need of services in the case of immigrants such groups unquestionably face great difficulty in communicating with health care providers. Health authorities have slowly established interpreter services, but even so the translation of language does not come to terms with the meanings which different cultural groups attach to symptoms, and to their concepts of illness and health (see Webb, 1982). Immigrants are, by and large, expected to seek advice and consumer services as though they had inherited the same values and concepts as the service providers. Since they are often unable to do so, they may reject advice and fail to comply with the instructions of doctors who come to see such patients as problems (see Forest and Sims, 1982 for a review of a special immigrant health advisory service).

Doyal (1979b) sees the same difficulties facing women who are confronted by a male dominated medical profession which expects women to behave in certain ways, and to perform roles as housewives and mothers which are then used to explain their search for medical help. Women are seen as 'more submissive, less aggressive, less competitive, more excitable in minor crises, more easily hurt, more emotional, more concerned about their appearance and less interested in maths and science' (Doyal, 1979b, 221). These expectations ensure doctors 'assume a neurotic basis for many of the problems presented by their women patients' (Doyal, 1979b, 226), a perspective reflected in drugs sales. In 1972 twice as many women as men received prescriptions for tranquillisers, hypnotics and anti-depressants.

One aspect of particular relevance to women's use of health services is their relative lack of mobility. Coupland (1982) has discussed this issue drawing attention to the problems women with children of pre-school age have in using health services when they do not have access to private cars. Coupland shows how private cars are not generally available to women during the day and highlights the disabling effects of costly public transport and vehicle designs which make it difficult for mothers with children and pushchairs to board. The Black Report was acutely aware of these issues and recommended that all NHS authorities should review the accessibility and facilities of the ante-natal and child health clinics in their areas. Evening and weekend opening times were suggested as one way of making clinics more accessible.

Unravelling the complexities of illness behaviour in attempts to explain observed patterns of utilisation is not easy. At the heart of the issue, however, is the meaning people attach to symptoms and this is clearly related to past experiences, and beliefs drawn from the specific social and cultural context in which the individual finds himself. People experiencing similar symptoms may well choose to act differently in response to them; some may accept them as a fact of life, a nagging nuisance to be lived with, others as a sudden divergence from their normal experience which requires immediate investigation and advice from the doctor. These responses are, however, also influenced by beliefs about the efficiency of seeking advice. Again past experience of medical services, established views of what doctors can achieve and the meaning individuals and their peers attach to being labelled as a patient influence the decision to seek medical advice. One element of this process which has not received as much attention as it should is that of lay beliefs about health, illness and medicine, and how families and social networks act as filters and exchanges in creating a cumulative experience which guides decisions to seek advice outside these contexts. It is known that the vast majority of symptoms do not result in a visit to a doctor, individuals either dealing with them or seeking preparatory medicine from chemists, or by taking advice from other lay persons. But frequently they do. The threshold at which this occurs is determined by the individual who undertakes a complex evaluation drawing on personal and peer experience, on personality and on the nature or the specific social and cultural context in which the individual lives. It is clear any policy which seeks to make NHS facilities more accessible to those who currently 'underuse' them must, if they are to succeed, come to terms with these socially derived constraints.

Professional Constraints

Illsley has specified the characteristics of a profession; central amongst them is:

the concept of autonomy – the legitimated freedom to define the nature of the work, to establish the content of training and the requirements for entry, to control the quality of work, to operate disciplinary procedures for infringement of membership rules, and to exclude the claims of others to compete in the same field. (Illsley, 1980, 60)

In the context of health care there are a number of professions providing services to patients which exhibit these characteristics – doctors, nurses, dentists, opticians and so on. Of these the doctors are undoubtedly the most powerful, since they are recognised as the head of the medical hierarchy, a position they jealously protect. Thus:

No doctor fails to recognise the necessity of co-operation with the nursing profession and with other medical workers. But this does not mean that the doctor should in any way hand over his control of the clinical decisions concerning the treatment of his patients to anyone else or to a group or a team. (Evidence from the British Medical Association to the Royal Commission on the NHS, quoted in Illsley, 1980, 63)

Doctors have achieved this position of dominance because we have entrusted responsibility for our health to them. We have done this because we believe the doctor knows best; he has, after all, undergone an extensive medical training which has provided him with the clinical expertise necessary to protect and restore our health. So entrenched is this belief that we have even given the doctor the power to decide if we are ill, for it is the doctor who legitimises our adoption of the sick role by the issue of medical certificates which excuse us from work and other social roles. Our willingness to accept this arrangement stems from the technical competence we expect doctors to possess (see Chapter 2). However, not everyone is prepared to tolerate the dominant position of doctors and medical autonomy is increasingly challenged (Elston, 1977). For instance, Kennedy (1980) argues that we have given the doctors too much power and he contends we must reassert our own claims, for health is too important an issue to be left entirely to doctors. Illich (1975) has gone further; he protests that the power of the medical profession has become a major threat to health as a consequence of iatrogenic disease and patient dependence on professionally administered health care of limited or unproven value. Nor does the power of the doctor stop at the application of medical therapies; they have successfully extended their influence over health policy and they have repeatedly ensured that their interests are protected in whatever arrangements governments have made for health care delivery. So who are these doctors? From where do they derive their power? How do they exercise it? And how does the patient as consumer attempt to control it? We shall attempt to answer these questions by examining the role of the medical profession in the British NHS and by comparing the influence of that role with that of the consumer as it is articulated through Community Health Councils (CHCs).

Doctors and Their Power

In the United Kingdom doctors derive the power to practice medicine from their inclusion on a register of recognised medical practitioners maintained by the General Medical Council (GMC). This is a statutory body created by the 1858 Medical Act and is ultimately responsible to the Crown as representative of the state. Strictly speaking it is, therefore, independent of the medical profession, though in practice the GMC is dominated by it. Members are drawn from the medical corporations like the Royal Colleges, from registered medical practitioners and from nominees of the Crown and the universities who, in any case, usually nominate members of the medical corporations (Forsyth, 1973). The chief role of the GMC is the setting and maintenance of professional standards which, though specified by the profession itself, are endorsed by the state. This is a crucial point, for it secures the professional autonomy of doctors. In so doing the claims of others to compete are excluded, so whilst it is not an offence to offer medical advice, it is an offence to pose as a 'registered practitioner' (MacKenzie, 1979). It also gives the profession the power to reproduce itself and to promote particular ideas about medicine and health. It is obviously important to understand, therefore, the social values potential doctors bring with them when they enter medical school, and the values and perspectives which medical education imparts if we are to make sense of the profession's exercise of power.

Medical education in Britain can be divided into two distinct phases: first, undergraduate pre-clinical and clinical training lasting for five years which takes place in medical schools attached to universities and teaching hospitals. Secondly, postgraduate training which is undertaken 'on the job' in hospitals, general practice and the community under the supervision and examination of, for instance, the Royal Colleges. The first phase may be regarded as a training intended to ensure competence to practice, the second phase being devoted to the development of specialist skills. Individual medical schools are responsible for the recruitment of students who follow a course of study approved by the General Medical Council, and inevitably they seek students who are likely to successfully complete the course. In practice this means recruiting students who have achieved 'A' level passes in science, usually in Physics, Chemistry, Mathematics and Biology. A number of studies have shown that these recruits have other social characteristics in common. Some 30 per cent of medical students in England come from private schools; in Oxford and Cambridge 50 per cent have parents who work in professional and managerial occupations (Social Class I); and as many as 30 per cent of all medical students are children of practising doctors (MacKenzie, 1979). Doyal (1979b) argues that these characteristics, as much as technical expertise, are responsible for the power and prestige doctors enjoy, but it is the technical expertise which doctors acquire in their medical education which is chiefly responsible for the perpetuation of a health care system based on medical intervention in disease processes. McKeown (1976) argues that the most important factor here is not the medical curriculum itself but the image of medicine projected in the teaching hospitals. Thus whilst the curriculum requires exposure to medical sociology, epidemiology, the mentally ill and handicapped, the greater part of medical education is taken up with the exposure of medical students to selective groups of acutely ill patients in teaching hospitals. The result according to McKeown is the production of medical school graduates who aspire to practice in the fields of medicine they have been exposed to in the teaching hospital, so we should not be surprised that 'a centre which excludes the mentally ill, the subnormal and many of the aged sick cannot expect to provide doctors who will care for them' (McKeown, 1976, 134). If McKeown is correct there is then something of a self perpetuating cycle which ensures the reproduction of certain approaches and perspectives to health care. Medical school selectors choose candidates in their own image and then transmit established medical values through the educational process which sustains those values and reproduces the distribution of power within the profession. For, tempting though it might be to think of the medical profession as a monolithic whole, the reality is of a profession in which power is unevenly distributed amongst the membership. The division of power is, however, a matter very much for the profession, for those who hold most power have it bestowed upon them by the prestige which the profession affords to particular aspects of practice. Stevens (1966) makes the point: 'As the Pharisee thanked God he was not as other men, do not Physicians preen themselves ever so slightly on not being Paediatricians, Psychiatrists, or Pathologists?' (Quoted in MacKenzie, 1979, 61).

It is not surprising, therefore, to find that the shortage specialties – that is those specialties where the demand for new consultants exceeds the supply of suitable candidates – are, according to the DHSS, mental illness, mental handicap, geriatrics, radiology, anaesthetics and the pathological specialties (Gt Britain, 1979b). Similarly it is consultants in these specialties who receive the fewest distinction awards – payments costing a total in 1979 of about £20 million per year, to consultants judged by their peers to have made a distinguished contribution to the NHS. According to the Royal Commission on the NHS about a third of all consultants hold distinction awards, though in England and Wales in 1977, '73% of consultants in thoracic surgery held an award, 64% of those in cardiology, 67% of those in neuro-surgery; while only 23% of those in geriatrics, 25% of those in mental health, and 26% in rheumatology and rehabilitation held awards' (Gt Britain, 1979b, 236).

This division of prestige within the profession is important as it adds a second dimension to the exercise of power by the profession. On the one hand there is the power the profession as a whole exercises in relation to the consumer and on the other there is the recognition that some groups in the profession can utilise this power more effectively than others because of the status they hold in a profession protected from external scrutiny by the overall autonomy which professional demarcation endorsed by the state brings.

Traditionally the most significant division of power within the profession was that between specialists (who usually worked in voluntary hospitals) and general practitioners. The division of prestige, power and wealth between these two groups is most aptly captured by Bernard Shaw in The Doctor's Dilemma. Compare Shaw's Sir Colenso Ridgeon, a distinguished physician who has just received his knighthood following his discovery of a cure for tuberculosis with Dr Blenkinsop, a general practitioner who is:

> clearly not a prosperous man. He is flabby and shabby, cheaply fed and cheaply clothed. He has the lines made by a conscience between his eyes, and the lines made by continual money worries all over his face, and hails his well-to-do colleagues as their contemporary and old hospital friend, though even in this he has to struggle with the diffidence of poverty and relegation to the poorer middle class. (Shaw, 1977, 108)

Forsyth (1973) has documented the recurring rivalries between these two traditions, but perhaps more important than these rivalries has been the ability of both specialists and general practitioners to protect their interests when the state has sought to extend its role in the provision of health care. Forsyth (1973) reports that the BMA spent £30,000 in campaigning to extract the best terms it could from Lloyd George's Insurance Act of 1911 which compulsorily insured low-paid workers for the service of a general practitioner. But this was merely a foretaste of bigger struggles to come, for the creation of the National Health Service by the post-war Labour Government is popularly characterised by the conflict between Aneurin Bevan and the various interest groups within the medical profession. This is not the place to rework the conflicts which arose during the setting up of the NHS, save to say that Bevan himself recognised he would have to make concessions to the medical profession if he was actually to bring the service into operation. Doyal (1979b) asserts that even though the major conflict was between the BMA and the Minister, it was the specialists who were conceded most, because they ensured freedom from local authority control, management by regional hospital boards and hospital management committees (on which they were well represented) and the right to private practice using designated 'pay-beds' in NHS hospitals. Moreover, Doyal (1979b) contends that since the specialists now had the backing of state finance for the hospitals in which they worked, the power of the consultant elite was consolidated and their status further guaranteed, with the result that acute hospital medicine came to dominate the NHS. This dominance has not gone unchallenged. Elston (1977) has explored how challenges to dominant groups within the profession, and indeed medical dominance itself, have grown since 1948. She identifies challenges arising from three sources: first, from changing patterns of morbidity and an ageing population which has forced governments and the profession to reconsider their priorities; secondly, from the growing strength of other health care workers who question the doctors' rights to dominance; and thirdly, from within the profession where junior hospital doctors amongst others have challenged the established leadership of the profession if not its overall autonomy. Significantly, though, Elston notes that whilst the appearance of these challenges 'indicates a shift in the balance of power away from the medical profession . . . For the most part, the attacks have been resisted' (Elston, 1977, 27). Principally this is because of the profession's stranglehold over decision making in the service, a position granted to them in 1948, and a position they have strengthened since.

This is not to argue that the medical profession should have no voice in management and planning. Bevan himself recognised the need for professional advice, though he made an important qualification to its exercise when he wrote, 'There is no alternative to self-government by the medical profession in all matters affecting the content of academic life, *although there is every justification in lay-co-operation in the economy in which it is carried out*' (Bevan, 1952, 90, emphasis added).

In 1948 Bevan sought to ensure such co-operation by the appointment of lay persons to NHS administrative bodies. The 1974 reorganisation of the NHS also appointed lay members to various administrative authorities, but also created Community Health Councils to represent the public voice. But how effective can these arrangements be in the face of professional medical power? Has there been satisfactory and sufficient lay co-operation? Or is Illsley (1980) correct in his assertion that professionally dominated managements have nullified that representation, and that CHCs were granted insufficient powers to counter such management?

Professional or Consumer Control?

Professional opinion is well represented in the NHS; consumer opinion is not. The 1946 National Health Service Act ensured that the professions, and especially the medical profession, would have an important role in the management and development of the service. On the other hand, patients as consumers have only been able to play a limited and often questioned role in the management process.

At a national level professional advice is provided by the Central Health Services Council (CHSC). Originally created in 1946 it is intended to provide health ministers with professional opinions about NHS policies and practice. Composed of representatives from the Royal Colleges and other professional organisations the CHSC has produced a number of influential reports, such as the Bonham Carter Report on the functions of the District General Hospital (Gt Britain, 1969). Health ministers are also advised by a series of Chief Officers who provide expert medical, nursing, pharmaceutical and dental opinions. The best known of these highly influential posts is probably the Chief Medical Officer, who has the equivalent civil service status of Permanent Secretary and with it direct access to the Secretary of State (Levitt, 1976). In addition, health ministers are able to obtain professional advice through the creation of special working groups (e.g. the Resource Allocation Working Party (Gt Britain, 1976b) and the Black Report (DHSS, 1980)) which are usually highly influential and on which the professions are heavily represented.

Extensive arrangements for the channelling of professional advice also exist at other levels in the NHS administrative hierarchy and professionals are key figures in the process of consensus management created by the 1974 reorganisation. The origin of these arrangements lies in a series of reports known as the 'Cogwheel Reports' prepared to identify ways of improving clinical involvement in NHS management. These reports recommended the creation of clinical divisions (groups of related specialties) in District General Hospitals which would be responsible to a Medical Executive Committee. After 1974 District Medical Committees replaced these, with the chairman and vice chairman being clinical members of the District Management Team. Area and Regional Medical Committees were also established and like their district counterparts are consulted on operational and strategic planning. So complex and time consuming has this process of advising become that the Royal Commission on the NHS recommended the Secretary of State to simplify professional advisory arrangements. Evidence to the Royal Commission from the Doctors' and Dentists' Review Body claimed that 95 per cent of consultants were members of at least one advisory committee, and up to one in six consultants members of five, with the result that, 'Representatives find themselves debating the same issues with very nearly the same people on different occasions' (Gt Britain, 1979b, 316).

The DHSS has subsequently undertaken a review of the advisory machinery. though the desirability and necessity of professional advice is still clearly recognised (DHSS, 1982). This underlines the influence which the professions have in the management of the NHS, in a wide variety of contexts. Thus, a recent report identified the following features of RHA activity as requiring medical advice:

long term planning of health care services; the arrangements for supra-regional and clinical services; the need and arrangements for regional and subregional services; the allocation of revenue and capital moneys between health authorities; setting of priorities for major capital investment in building and equipment; the deployment of hospital medical and dental manpower; the provision of appropriate resources for undergraduate teaching and research; the encouragement of clinical and health service research; the development of policy for the provision of postgraduate medical and dental education; the provision of a careers advisory service for doctors throughout the service. (BMJ, 1982, 64)

Advice, of course, need not be heeded; so who receives and acts upon it? The answer, following the 1982 restructuring of the NHS, is the professional officers who make up the District Managements and the Regional Team of Officers (RTO) who, in turn, advise the members of the District and Regional Health Authorities. RTOs consist of a Medical Officer, Nursing Officer, Administrator, Treasurer and Works Officer who forward proposals to meetings of the full Regional Health Authority which are held throughout the year. Regional Officers are responsible for the daily activities of the RHA, and their collective opinion carries great weight with the Chairman (who works closely with the RTO) and the members of the RHA, who are appointed by the Secretary of State following discussion with the health professions, local authorities, universities, health related trade unions and voluntary bodies (Levitt, 1976).

Similar relationships between members and officers existed at the level of the Area Health Authorities (abolished on 31 March 1982) and are likely to recur in the superseding District Health Authorities. Key officers in these new authorities are the District Medical Officer, District Nursing Officer, District Administrator, District Finance Officer, together with a GP and a consultant representative with whom responsibility for planning and managing local health services now rests. It is pertinent to add here, that professional opinion discussed in this context must include that of the professional administrator. Apart from the civil servants who staff the DHSS, the most senior of whom carry great influence since they are often in post before and after many ministers, there is a whole network of administrators who are members of, and some who support, officer teams at regional and district level. These too carry considerable influence in the day-to-day decisions which take place in the NHS, having grown in number by 28 per cent between 1973 and 1977 (Gt Britain, 1979b). As with other professions they have their own professional organisation – the Institute of Health Service Administrators – and there is no doubt they are a potent force in the NHS.

From the above it is evident that there are a number of routes and levels whereby professional, and notably medical, opinion finds its way into NHS decision making. There can be no doubt about the power of this opinion to guide NHS planning. As Professor Perrin's report on the management of financial resources in the NHS pointed out:

> Clinicians are a highly intelligent and articulate group, and are invariably respected by teams of officers. Indeed, they may be so formidable as to dominate decisions about resource use sometimes using the doctrine of 'clinical autonomy'. This autonomy, however, cannot in a public service be absolute . . . the exercise of clinical autonomy ought not to be allowed to extend to a veto on reallocation of beds to cope with changes in need. We came across a flagrant example of such a veto, and elsewhere comments were made about the reluctance of authorities to take prompt action to reallocate beds. (Gt Britain, 1978, 45–6)

Compare this with the position of the consumer, who has no direct say in the appointment of persons to represent his interests. As Illsley points out, 'of all the major services, only the health service is not directly administered by elected representatives at national or local government level' (Illsley, 1980, 94). Local authorities have argued that health care should be one of their responsibilities (they did in any case operate a large number of hospitals before 1948 and indeed provided some community health services between 1948 and

1974) for it would unite responsibilities for health with those for
social services and subject them to local electoral control. In
evidence to the Royal Commission the Association of County
Councils said:

> the most crucial decision is whether the service should be
> returned to the public. Such a step, with the agreement of the
> professions, would go a long way to ensure a health service which
> belongs locally and to which people feel committed. (Gt Britain,
> 1979b, 264)

The Commission was not convinced by these arguments.
Absence of a regional tier of local government, concern that local
authorities might do no better than NHS authorities and worries
that local authorities could not cope with the administration of
another major service, led the Commisssion to reject the idea.
Representation of the public's viewpoint rests principally,
therefore, with the lay members of the RHAs and DHAs and with
the Community Health Councils created in 1974. Additionally the
public has certain legally protected rights (see Stimson and
Stimson, 1978 for an excellent presentation of these); there are
complaints procedures covering all branches of professional
medical practice, a Patients Association and a Health Service
Commissioner – the NHS's Ombudsman. Of these CHCs are the
most important counterbalance to the professionalised management of the NHS. They were created to represent the views of the
consumer at the local level, and are intended to interact with health
authorities in the development of local health services, yet
significantly they are not elected councils. In common with NHS
authorities members are appointed by individual RHAs according
to guidelines issued by the DHSS, a chairman is elected by the
membership, and a secretary appointed following open competition. Membership guidelines have recently been reviewed as part of
the 1982 restructuring, though not before the whole future of the
CHCs was called into question by the government, despite advice
from the Royal Commission that they should be supported. In a
circular issued in December 1981 the Secretary of State announced,
'that CHCs were to be retained but subject to a review at a date in
the near future when their operation in conjunction with locally
based DHAs could be assessed' (DHSS, 1981b). One observer
claims the reason for this policy is 'the suspicion that the councils

are a Trojan horse left by Labour, waiting to discharge bands of hostile left wing saboteurs into the NHS' (Deitch, 1982, 117). Do CHCs represent such a threat? What have they, or haven't they, achieved and how has the NHS and its professionals reacted to them?

There are no all embracing answers to these questions; the work, and reactions to the work of CHCs, have been extremely varied. In some places CHCs have been aggressive in opposing health authority plans, in others they have been submissive. Some authorities have taken CHCs into their confidence and encouraged participation in planning, whilst others have considered such involvement as wholly improper, forcefully rejecting any such aspirations the CHC may have. In part the variable experience of community representation can be attributed to a lack of clarity of CHC objectives – not only amongst CHC members, but also in NHS management, including the DHSS who, having invented such a body, 'then decided what it should do. As we worked on the CHCs, we found more things for them to do' (Klein and Lewis, 1976, 15). Following the 1982 restructuring certain rights have been reaffirmed. CHCs have the right to make visits to health service institutions (though not GP premises), and can send an observer with speaking but no voting rights to meetings of the DHA. The CHC can refer plans to change the use or close health service buildings to the Secretary of State and will also publish an annual report to which the local DHA must reply, in addition to arranging a joint meeting with the CHC at least once a year. DHAs are expected to 'make every effort to consult CHCs in good time on all matters of interest to them' (DHSS, 1981b). Experience between 1974 and 1981 suggests that there is some dispute about matters of interest. In particular, CHCs complained to the Royal Commission that they often lack resources to investigate matters of interest or prepare counter proposals, and that in the case of Family Practitioner Committees (administrators of primary care services) they found considerable difficulties in gaining information about any plans for local services. Although some FPCs admit CHC observers to their meetings not all do so, and whilst the Secretary of State has encouraged more to adopt this practice he has not made it a requirement. At the same time the Secretary of State will not be making any additional resources available to CHCs.

Effectively representing the public's viewpoint may thus become more difficult for CHCs. But consumer participation in management and planning is not just dependent upon adequate resources; more fundamentally it is dependent upon the relationship of the consumer representatives to the management authorities and, in the case of CHCs, they have very firmly been kept at arms length and have been granted no executive powers (Doyal, 1979b). Herein is the major difference between professional and consumer relations within the NHS. The former are an integral feature of the planning and management process at all levels of the administrative hierarchy and in all spheres of NHS activity, with access to and command over considerable resources, with the unquestioned right to give advice to the unelected laymen who sit on health authorities. The latter are collections of lay individuals with limited time, often with other representative commitments, few resources and constrained rights of action but, most importantly, external to the decision making processes.

There is something of a dilemma here, however. So long as CHCs remain outside the formal planning machinery they preserve their independent status and the right to criticise, but the price they pay is to have little say in the development of policy, for the authorities can simply ignore CHC arguments and nullify alternative proposals by starving CHCs of information. On the other hand, if CHCs become integrated they cease to be identifiable as specific representatives of consumer opinion, becoming more like existing lay members of various administrative authorities. In other words, there is a danger that the CHCs may become co-opted – additional agencies of bureaucratic practice and control. There is no easy resolution to this dilemma, since if CHCs fully participate with management they may lose the (tacit) support of the community they represent, especially at times when services are being cut back. If they remain outside the decision making process, administrations are able to dignify their policies by claiming on some occasions that CHCs support them, and on others when opposed, they are able to claim to have at least consulted the community's representatives (Heller, 1978).

Creating this dilemma is, as Heller (1978, 61) comments, 'in the very best interests of the major power groups', who in retaining the control decide when and on what issues the CHCs should be consulted. Moreover, if they find themselves in opposition the authorities can always question the representativeness of the CHC. This is not the case with professional advice which is regarded as vitally important. This is not to argue that CHCs are wholly ineffective or professional advice wholly unwelcome, but to emphasise that the balance of power rests with the professions, and that for any policy to succeed it must have their support. CHCs have made, and will no doubt continue to make, a contribution to the NHS, but it seems inevitable that the realm of their activities and their impact will continue to be circumscribed by professionally dominated managements.

Conclusion

In this chapter we have tried to show how the utilisation and planning of health services is not simply a matter of measurement, quantification and allocation. On the contrary we have tried to point out how historical, behavioural and political influences create a reality which is often markedly different from a specified optimum. We have structured our discussion and presented examples which may suggest these constraints are not linked, but in truth the distinctions we have drawn are more apparent than real. We can illustrate this by referring to the continuing debate on primary care in inner London which draws elements of our earlier presentations together.

As we have seen, proposals exist to reduce the number of hospital beds in inner London. There is considerable concern that implementing these proposals may place additional burdens on unsatisfactory primary care services, especially in light of the Royal Commission's conclusion that 'the NHS is failing dismally to provide an adequate primary care service in the inner cities' (Gt Britain, 1979b, 89). This observation is particularly true of inner London. Subsequently the problems of primary care have been the subject of a special report by the London Health Planning Consortium ('The Acheson Report', DHSS, 1981a).

The report identifies two particular problems; access to services and the quality of premises. Evidence gathered by the LHPC showed that many general practices were difficult to contact outside daytime surgery hours as well as during the night. Many premises were also considered to be unsuitable for modern general practice with its emphasis on team work. These deficiencies are especially significant given the nature of inner London's population with its disproportionate numbers of elderly, immigrants and transients. Here is a classic example of Hart's (1971) 'inverse care law' – a socially disadvantaged population being served by primary care services which are considered inadequate. Some patients overcome the problem of GP access by attending casualty departments in nearby hospitals (often teaching hospitals) which thus become surrogate sources of primary care. Some people are probably deterred from seeking care when they ought to, and under-utilisation of preventive services almost certainly results.

Ironically, though, inner London is considered to be 'over doctored' according to the criteria adopted by the Medical Practices Committee (MPC) (described in Chapter 4) which employ mean patient list size as the arbiter of need. Patient list sizes in London are small for two reasons. First, some GPs keep their lists small so that they can devote a large part of their time to private practice whilst at the same time collecting basic practice allowances (worth in 1982, £5,320 per annum) from the NHS when their patient list exceeds 1,000. Secondly, list sizes are small because the population of inner London has declined rapidly, but the stock of doctors has not. Average list size in inner London in 1979 was 2,151 compared with 2,275 persons in outer London, but 17 per cent of doctors had lists smaller than 1,500 persons compared with 10 per cent in outer London and 7 per cent in England and Wales (DHSS, 1981a, 25). Likewise a disproportionate number of inner London GPs are aged over 65, and a disproportionate number also work alone rather than in group practices. Eighteen per cent of inner London GPs were aged 65 compared with 6 per cent nationally, and 59 per cent of GPs worked alone compared with 28 per cent nationally. Single handed practice, especially by doctors over 65, was considered by the Acheson Report to be undesirable and as a remedy the creation of group practices and primary health care teams was advocated. Other recommendations in the report sought to improve the quality of practice premises, and to create practice vacancies by instituting a policy of retirement which would reduce the number of elderly doctors and increase average list size.

Achieving these ends however, is no easy matter because the NHS has only limited control over general practitioners who negotiated independent contractor status when the NHS was set up. Practically, this means GPs only contract to provide services; they are not NHS employees, and they retain a strong voice in the administration of primary care. Independent contractor status is jealously guarded, and the outcome of Acheson's strategy for improvement will depend upon the ability of the government to negotiate amendments to GP conditions. If GPs are unwilling to accept new arrangements then it is unlikely the proposals will be implemented, since GPs cannot simply be coerced into adopting particular practices. The existing lack of control over GPs is aptly demonstrated in the case of surgery standards. Family Practitioner Committees should ensure that standards are maintained, GPs being expected to provide 'proper and sufficient' accommodation. Thus, 'the GP has to obtain the agreement of the FPC that his premises meet these requirements but no minimum standards . . . are laid down; and consent may not be unreasonably withheld' (DHSS, 1981a, 33).

We have here all elements of the constraints we have described operating simultaneously. Historical factors have contributed to the creation of the problem; in consequence the population finds it difficult to gain access to doctors, even though numerically they are plentiful, and the scope for change is constrained by the contractual relationship between GPs and the NHS – a relationship which GPs fashioned and protect. This example and, indeed, all of the examples we have used to illustrate the operation of historical, social and professional constraints relate to specific contexts. Most of our examples have been drawn from Britain, and as such are an articulation of their operation in an advanced capitalist state with a publicly financed health service. Undoubtedly the constraints we have described operate in other socioeconomic contexts, but the form they take and their particular significance is inevitably related to the nature of that context. Thus, we can see the nature of society itself as an additional, important constraint, and it is to this issue we now turn.

In the previous chapter, we examined the nature of the social-behavioural and professional-institutional constraints that affect the provision of health care. We now wish to reintroduce some of the themes discussed in Chapters 1 and 2, especially the relationship between medicine, health and society. We contend that the nature of health care provision is dependent upon the nature of the containing society and the conceptions of health prevailing in that society. In other words, societal constraints, including the cultural definition of and response to health and illness, shape and influence the nature of health care and the development of health care policy. Thus, these constraints affect what states do and do not do. As Higgins (1981, 17) argues, it is just as important to know 'what governments *do not* do in relation to certain areas of need, as well as what they *do*'. The United States, for example, has failed to introduce a national system of health insurance for all groups, although it is a policy option. It is possible to argue that it is the nature of American society in terms of its dominant interests and its cultural conception of health and health care in terms of the ideology of individual responsibility and self-help that explains such a 'non-decision'.

We shall examine three instances of the impact of societal constraints on different types of society. The first is based on the somewhat artificial distinction between traditional and modern systems of health care. We shall examine the relationships between conception of health and nature of care in small-scale societies, where environment and social structure are also important mediating factors. Secondly, we shall examine some advanced industrial nations and suggest that the nature of health care is again dependent on the conception of health but that, more importantly, this is in turn based on a particular view of man and society. Thirdly, we intend to take such models further by isolating some state socialist societies in which the rural sector is still extremely important and which because of resource limitations emphasise preventative rather than curative medicine. As will be readily discerned, such distinctions may be more apparent than real and the

183

categories merge into one another. Thus, traditional systems are being modified by the penetration of allopathic medicine, while in practice in the third category curative medicine is important, especially with the gathering of industrial and technological change.

Conceptions of Health and the Nature of Health Care in Traditional Systems

Traditional systems of care are still widespread. Scarpa (1981) suggests that what he calls pre-scientific medicines treat up to 80 per cent of the world's population. He goes on to suggest that this is in large part a consequence of the connection between such medicines and religion. We should add that it is not simply religion *per se* that is important. Religion presents a way of ordering and understanding both the social and natural worlds. Thus conceptions of illness, health and treatment, derived from religion, fit with the rest of the social framework which is also so based. Perhaps the classic example of the defining principles of religion on sickness is the study of the Zande of the Southern Sudan by Evans-Pritchard (1937). Amongst the Zande, a widespread belief is that illness is caused by witches who possess a special kind of power. Indeed most misfortunes are attributed to witchcraft. The Zande in effect maintain the pretence that people should never die; mortality is ascribed to witchcraft. Thus if a person falls ill, witchcraft is suspected and an oracle is consulted. By using an oracular technique – the manipulation of a special rubbing board, the administration of a strychnine poison to fowls – the sick person attempts to identify his attacker. While such a system is open to manipulation, oracular verdicts are always justifiable even in terms of suggesting that the oracle was bewitched if an incorrect judgement later becomes apparent. Thus the ultimate effect is reinforcement of faith in the efficacy of mystical power and the oracle. This has societal importance because Zande oracles, like the political system, are hierarchically organised with the most powerful being operated by the chiefs themselves. The system of oracles is, therefore, consistent with and may be said to support the political system with which it is associated. Notions of illness and treatment are inextricably linked to the social order.

The Zande are not completely mystical, but witchcraft and oracles remain important today among this Christianised society.

They accept other explanations, but witchcraft and mystical beliefs are invoked as causal explanations of irregularities. Thus leprosy is seen as being caused by incest – the neglect or transgression of taboos. Witchcraft also supplies a deeper cause – with structural implications – than that supplied by ordinary everyday reasoning. Indeed, Evans-Pritchard suggests that amongst the Zande, every illness supposes witchcraft and emnity, implying that apparently supernatural explanations predominate. In this he is supported by Assimeng who argues that in parts of Ghana 'Illness is seen as dirt, pollution and a danger, seen as an extension of *honhom fi* (evil spirit). It is in this regard that ill health is not seen in mere rationalistic terms' (quoted in Fosu, 1981, 475). Other studies have suggested that there is a continuum between natural and supernatural causes. Among the Ogori in Nigeria, some events – the death of a young adult – are regarded as unnatural, others are regarded as serious and mysterious but curable, while others – malaria – are seen as natural. These naturally caused diseases are in fact seen as part of the normal order. The definitions of a disease as 'normal' with the implication that treatment is unnecessary and unhelpful is not reserved for small-scale societies. Koos (1964) points out that lower back pain is regarded as normal in the American city of 'Regionville' and is thus not appropriate for medical attention.

Fosu (1981), in a study of disease classification in rural Ghana, suggests that diagnosis of cause is the most important aspect of such classification and that there are three major types of causes. First, there are diseases caused by natural agents (worms, insects, unhealthy environments, the malfunctioning of specific organs); secondly, those caused by supernatural agents (identified by the breaking of taboos or social-religious injunctions); and thirdly, those caused by both. If supernatural causes are identified, good health is dependent on the observing of moral and religious norms. In this instance, disease is seen as retribution, although witches and demons may inflict disease to disrupt group harmony. In this instance, we see that the supernatural explanation of disease serves a social purpose. The religious belief behind such explanation and the ritual atonement necessary to regain health act as factors ensuring social cohesion. Thus, among the Gimi of New Guinea, illness has communal meaning and its treatment is directed at solving social conflicts (Glick, 1967), while in the village culture of Sri Lanka, mental illness and the rituals for its treatment serve to reintegrate families and re-establish their social boundaries during periods of stress (Waxler, 1977).

In Fosu's study of Berekuso 56 per cent of all reported disease was attributed to natural causes (impurities in the blood, head and stomach; accidents; insect bites), 13.5 per cent to supernatural causes and the remainder to a combination of the two. It was also found that the explanation of the disease influenced the type of care sought. Of the 'natural' diseases 53 per cent were treated at clinics and 17 per cent by the individual or his family. Interestingly, nearly 25 per cent received no treatment. Of the 'supernatural' diseases only 15 per cent were treated in the clinics while 31 per cent received treatment from traditional medical practitioners and 42 per cent from self or family. Only 11.5 per cent received no treatment. Thus we can see that there appears to be a close relationship between the conception and explanation of ill-health and the nature of health care sought. Such a view is supported by Alland's (1970) study of the Ivory Coast. This suggests that the definition of disease determines care. Thus, if the illness is seen as mild or common, it is self-treated using Abron, Moslem or Western medications. If it is viewed as serious the advice of a non-professional specialist, that is a person with a knowledge of herbal remedies, is sought. If it is regarded as severe, then a professional is consulted to seek the social or supernatural cause and to prescribe the correct medications and supplications. Alland points out that any of these steps can involve referral to the government or mission dispensaries.

Traditional systems of care are thus intertwined with Western practices. In fact, it is often the case that modern medicine is seen as treating the symptoms of a disease while the traditional exposes its causes. Further, from their study of the North Solomons, Hamnett and Connell (1981) suggest that if illnesses are treatable by modern medicine they are transferred to a category of lesser importance. Western medicine is thus grafted onto a pre-existing system that relates traditional medical practices to society and polity and as such it is a mere adjunct to the system that locates the individual and his health problems in the entire social fabric. We should be careful of making too broad a generalisation because, in a recent study of the Ivory Coast, Lasker (1981) suggests that the choice of therapy is dependent on accessibility rather than the characteristics of the individual patient. Further, the choice of Western medicine is not inhibited by unscientific attitudes but by political and economic constraints on the usefulness of such medicine and by the availability of attractive alternatives. While Lasker's conclusions do not seriously question the relationship between form of medicine and social order they should make us examine any such relationship critically.

In fact it would be incorrect to suggest that Western medicine and models of society (see below) have not affected the nature of health care in less developed societies. Croizier (1970) suggests that the waxing and waning of the fortunes of the traditional and modern systems of health care is closely tied to changes in power relations in the societies concerned. Health care is still related to the nature of society, but its provision is determined by the association of dominant groupings to the various types of medical practice. Leslie (1974), for example, demonstrates that the similar traditional systems of China, Japan and India had different fates. In Japan, the ruling elite adopted scientific medicine as the legally sanctioned system, while in China there was, and is, political support for the incorporation of the traditional into an essentially Western system. In India, there is a dualistic system, the traditional and modern serve and divide different castes and classes, despite the fact that most social groupings prefer Western medicine (see Banerji, 1975). Frankenberg (1981), however, argues that Western medicine has helped to legitimise and reinforce capitalist state power in both the colonial and independence eras in India. Western medicine is the bearer of an urban, male, technological, hospital-based, cosmopolitan, curative, individualistic world-view. Traditional medicine, embedded in family, Moslem Unani and Hindu Ayurvedic practices, is incapable of providing a coherent overarching view, even of disease, because of the limits of the economic structures in which it is embedded. This is not to say that these traditional forms are unimportant. Tabor (1981) points to the continuing role of Ayurvedic medicine in Southern Gujarat. Such medicine places great emphasis on digestion which is seen as 'gastric fire', the mediating factor between the internal (bodily) and external environments. Faulty digestion is manifested in 'unripe' stools which contain *āma*, the substance by which symptoms are recognised. Thus, unwholesome diet is seen as the chief cause of disease which is treated by fasting or diet to stimulate the gastric fire or therapy to purge the body of *āma*.

This example not only shows the relationship between traditional and modern systems but also demonstrates yet again the relationship between health care, social thought and social order. Parallels are drawn between ripe and unripe foodjuices and stools and perfect and imperfect foods. Analogies are also made between digestive and sacred fire. Thus 'wholesome' food and diet are equated with pure food in both a ritual and non-ritual sense. Health is associated with living according to the precepts of Hinduism, which was the moral order underpinning Indian society. Today though there is evidence of 'medical pluralism' and the adaptation of the Ayurvedic system to Western medicine. The diagnoses of allopathic doctors are accepted and patients are treated on the basis of these as well as by traditional therapies.

Dominant interests have, then, the power to shape the health care system and to determine which modes of care will be emphasised. Such power can even affect the definition of illness. Thus, administrators and churchmen taught the Navaho Indians in the American Southwest to see drinking as an illness, enabling medicine to become an agent of social control (see Kunitz and Levy, 1974). Zaire, however, provides an example of the relationship between traditional and modern being influenced not so much by consumer interests and preferences but by the colonial and then post-independence elites. The Belgian colonial administration harassed traditional healers, seeing the diagnoses and divinations of prophet-seers as conflict-arousing as they tended to politicise the indigenous population. The practice of witchcraft became a punishable offence. Herbal treatments were tolerated, but the biggest advance was made by Western medicine which became regulated and codified, and hence part of the established order. Janzen (1978) notes that with the establishment of African power in 1960, Western medicine still predominated but traditional healers became more important. It was, however, with the rise of Mobutu in Zaire that traditional medicine finally re-emerged. This re-emergence can be seen as part of the legitimisation of the Mobutu regime for it stressed de-Europeanisation, independence and African cultural authenticity. The traditional leader (*banganga*) became again a valid part of African heritage, a main prop for Mobutu. The *banganga*, therefore, became organised and codified as had Western doctors under the Belgians.

In many countries Western medicine, as representative of the global force of capitalist economic and political relations, has shaped the entire health care system. Thus Goldstein and Donaldson (1979) show how the Western training of Thai doctors has led to such training dominating the nature of medical care in Thailand. In Ghana, over 75 per cent of health expenditure is on curative services in general and hospitals in particular. Over 75 per cent of all doctors are in urban areas which contain only 23 per cent of the population. Of greater benefit to the health of the population would be a modern sanitary system. But the hospital sector is prestigious and Western-trained doctors bring back the ideas and ideals of scientific medicine to their own societies. This is a subtle, almost unintended influence but one which helps divert attention from the most pressing health needs. In fact, in most parts of the world these are visible and monotonous – malaria, diarrhoea, pneumonia, bilharziasis, etc. Indeed, these health problems demonstrate most depressingly the relationship between health care (or the lack of it) and societal changes and developments which, we would argue, reflect and serve the interests of dominant groupings and nations in the world economy and polity. We believe that our comment is supported by the more bland language of the World Health Organization (1980, 2–3) which points out that:

> The most important social trends are reflected in the still low, and in some areas worsening, nutritional levels of the bulk of the population. The employment situation, including access to land, has not improved in many countries and is partly, but not primarily affected by continuing high rates of population growth. The decline of rural life in many countries has led to unacceptable rates of urbanization and social and health problems on a mass scale in the world's cities and large towns. In addition, even when economic growth has taken place, the distribution of resultant benefits has sometimes been such as to widen the social and health gap within countries . . . continuing poverty . . . is at the root of the world's most pressing health problems.

While this book is not the place to describe the interdependence of the world-system (see Wallerstein, 1974), the WHO statement is an eloquent pointer to the relationships between health and society, relationships that will also inform our discussion of health care in advanced industrial societies.

Health Care Policies in Advanced Industrial Societies

It is of course a truism to suggest that advanced industrial societies do not have such a close relationship with their environments as do small-scale and pre-industrial societies. Such relationships, and even those between members of groups, are mediated by technological and institutional structures. While we still see associations between illness, health care and society, they are likely to be of a different kind. We shall suggest that not only key institutions – the state, the church, the family – but also particular models of man and of society influence the nature of health care. Thus we shall implicitly re-examine some of the themes outlined in Chapter 2. As also in that chapter, we do not follow the logic of industrialism argument which suggests that the welfare state is an inevitable outcome of industrial development. All societies have experienced different histories, enjoyed different environments, created different institutions and developed different views of the world, all phenomena that shape the nature of health care provision. These phenomena result in different principles being adopted to allocate resources to health care and then to distribute those resources between groups and territories. These principles may vary over time within societies but it is possible to suggest that they are predicated on particular models of man and society. It should also be noted that these principles often have to be adapted to the exigencies of economic and political pressure from within and without. Thus, for example, health care in the Soviet Union 'is regarded as one of the most fundamental functions of the state on a par with guaranteeing the citizens the right to work, rest and education' (Lisitsin, 1972, 46). It is thus the state's role to fulfil the function of implementing fundamental social rights (see Chapter 2) on the principle 'to each according to his need'. But from its inception, the administration of Soviet medical care was affected by political conflicts such as those between the professional physicians and powerful trade union committees (see Kaser, 1976). In contrast, health care in the United States is mainly provided by private enterprise, based on a view of society dominated by market principles and individual, self-interested decision-makers. Health care provided by the state is founded on the ideas of less eligibility and rigid means-testing and rationing, that is on civil rights (see Chapter 2).

As we shall see, the case of Britain is again different. The provision of health care facilities may be regarded both as a way of integrating the working class and as a moral obligation to care for citizens. Such moral-integrative principles operate in other societies but in rather different ways. Much indeed depends on conceptions of health and welfare, the institutional sources of such conceptions and the interests served by the moral integration. Thus, in Japan, many aspects of social and health policy were shaped by Buddhism with its emphasis on the virtues of thoughtfulness, sympathy, kindness, pity and benevolence (see Sugimoto, 1968; Higgins, 1981). A close relationship was gradually established between government, religion and social policy. Western influence meant the introduction of more advanced medical techniques and the ruling elite adopted scientific medicine as the legitimised system of health care. In fact, such treatments are usually provided for the employee by his employer. Mishra (1977, 96) suggests that this provision 'forms a part of the traditional (feudal) paternalistic relation (familism) between the employee and the employer'. Health care provision thus helps to legitimise the nature of work relationships. Scientific medicine is, in the main, part of occupational welfare (see below) demonstrating a close association between health and the social order, an association that is mediated by a model of society which emphasises the moral responsibility of employer to worker. It so happens that that responsibility also serves the interests of employers, helping to ensure a loyal, healthy workforce. The pattern of health care provision in Japan is not that straightforward, because Buddhism is still influential. Sugimoto estimates that about 70 per cent of social work in Japan is still under religious auspices. Thus a historical model of man still has an important impact on the everyday lives of many Japanese.

One further example will suffice to show the variations in the application of moral-integrative principles. Higgins (1981) demonstrates the conservative role of religion, namely the Catholic church, in the development of social and health policies in Eire. This role in policy is simply in microcosm the effect of the church on society as a whole. Indeed, the church has had a great influence on the state and part of the Irish constitution, largely based on Catholic teaching, encourages citizens to provide for their needs through their occupations rather than seeking external support. The state, however, undertakes to provide for those whom it deems are the weaker members of the community, particularly women and children. But the state's role is mainly to supervise and assist the harmonisation of group interests, because Catholic teaching argued that different social groupings would co-operate, without the need of state intervention, to create an effective and compassionate society in which the family was the most important social unit. In the immediate post-1945 period, the power of the church was such that it successfully opposed health legislation, particularly that to provide free health education and medical care for mothers and children up to the age of 16 (see Coman, 1977). In essence, the fear was that doctors would provide an alternative model of society, as they would be able to instruct in sexual, marital and even behavioural matters. In more recent years, though, greater prosperity, public expenditure, the broadening of Irish horizons and the liberalisation of the church itself have meant that the church is but one influence on health policy. In any event, this example demonstrates how health care provision can be seen as central in the struggle between two powerful world-views – church and state – both with their models of man and both working to enhance their own interests, religious domination and electoral success. As we have said, health care policy in advanced industrial nations is the outcome of history, ideology and conflict. We wish to examine this contention, as well as make an assessment of the limits placed on policy options by existing practices, in our major examples – Britain and Australia.

Health Care Provision in the Advanced Capitalist State

In examining Britain and Australia, we shall argue that the major constraint on the provision of health care is the need to ensure the reproduction of the capitalist state. We adopt the view that in advanced capitalist nations health care largely remains the responsibility of the individual, governments only intervening to provide a medical care safety net as a means of ensuring social integration. In this context health status is not always seen as a collective responsibility although citizenship does ensure some social rights (see Chapter 2). Instead the individual's health is regarded as a private matter between him/herself and the doctor. The practice of medicine becomes, therefore, another technical skill wielded by individual professionals, in a personal, quasi-private (and sometimes pecuniary) relationship with patients treated as largely divorced from the economy and society generating unhealthy environments and their own ill-health.

Indeed. the biomedical perspective of doctors – their model of man – is consonant with society conceived as a collection of self-interested, but enlightened, individuals – a model of society (see Chapters 1 and 2).

This does not mean there are not conflicts over health policy or spatial resource allocations. Policies and allocations change as the relations between power groups – including capital and labour – change, and more specifically as the relations between political interests change. 'Political interest' must be regarded broadly; it does not simply refer to party political, but institutional, economic and professional interests as well (see Chapter 5). But the crucial point is that these conflicts are subjugated by the needs of the economy and society in which the state is located. In advanced capitalist nations this means that the state's involvement in the provision of medical care is principally to ensure the reproduction of the state and the capitalist mode of production. It may also be possible to argue that medical provision in state socialist societies ensures the reproduction of the 'socialist' state.

To illustrate this perspective we have selected Australia and Britain as examples. We explore how in the case of Australia the influence of the state widened during the early 1970s as it attempted to take great responsibility for health care, but how since the mid-1970s the health care system has progressively been returned to the private sector in the belief that this releases funds from unproductive social expenditure. In the case of Britain we examine the expenditure priorities of the NHS and show how these have not been realised, again as the consequence of decisions to reduce unproductive social expenditure. Thus we assert that the differences between the British and Australian examples are but differences of degree.

Health Care in Australia: Public or Private Provision?

Dewdney (1972) suggests that the debate over health care provision and the role of the state in Australia can be divided into four phases. The first, before 1941, was characterised by unsuccessful attempts to introduce compulsory insurance schemes. The second, between 1941 and 1949, saw attempts (unsuccessfully) to introduce a mass comprehensive NHS type of service. In contrast the third phase, lasting between 1950 and 1969, was a period when health care provision was dominated by private insurance schemes; and the fourth, from the beginning of the 1970s, associated with reforms to

Table 6.1: Changes in Health Care Plans in Australia

1953–72	During this period, health insurance was generally provided by friendly societies which were subsidised by the Commonwealth Government. The basis of the scheme was that patients attended and paid the doctor of their choice, then got a refund from one of the government-approved and subsidised friendly society funds.
1972	The Australian Labour Government came to power.
1972–5	The existing system continued whilst the Labour Government formulated its government-paid health insurance system called Medibank.
1 July 1975	The Medibank health insurance scheme was introduced and covered everyone in Australia for at least 85% of doctors' charges up to the level of the government-approved fees and provided free public hospital accommodation and treatment by a hospital-appointed doctor. Medibank was financed from general taxation revenue and doctors who chose could send their bills directly to Medibank under a bulk-billing system.
December 1975	The Liberal Government came to power and established a Medibank Review Committee.
1 October 1976	The Liberal Government introduced a number of changes to the Medibank system. These were: (a) the imposition of a 2.5% levy on incomes to a maximum of A$300[a] per year – family – and $150 per year – single – instead of financing Medibank from general tax revenue; (b) allowing people the choice of opting out of Medibank and the 2.5% levy, provided they were covered for medical and hospital costs with a private health insurance fund; and (c) the establishment of Medibank Private as a health insurance fund in competition with other funds.
1 July 1978	Medical benefits were reduced from 85% to 75% of government-approved fees.
1 November 1978	Medibank Standard was abolished; health insurance ceased to become compulsory; and a Commonwealth medical benefit was introduced. This Commonwealth medical benefit paid 40% of government-approved fees for everyone, whether they were privately insured or not. In addition, there were no bed charges for public ward accommodation in public hospitals.
1 September 1979	The Commonwealth medical benefit was altered by abolishing the 40% universal rebate and replacing it with a system which rebated all amounts of doctors' charges at the government-approved level which exceeded $20 per service. Public hospital charges were also increased by 25%.
1 July 1981	Contributions paid to registered health funds for cover up to the level of 'basic health insurance' were allowed an income tax rebate of 32 cents in the dollar.

| 1 September 1981 | The Commonwealth medical benefit was abolished and replaced by a government subsidy amounting to 30% of doctors' charges up to the government-approved level, which would be paid directly to registered health funds and, therefore, only available to those persons insured by a registered health fund. Public hospital charges were increased by 60%, and the basic health fund benefit was increased from 75% to 85% of doctors' charges approved by the Government. |

Note:
a. A$1.60: £1.00 as at 17 August 1981.
Source: O'Sullivan (1981), p. 7.

the arrangements developed between 1950 and 1969. No doubt if Dewdney was writing his book today he would identify further stages, notable amongst which would be the period between 1 July 1975 and 1 October 1976 when the Labour Government introduced a government-paid health insurance scheme called Medibank, predicated on the recognition of a needs nexus and the view of the social system as an interdependent moral community. During this phase, state participation in the provision of health care reached its maximum extent, since the period after October 1 1976 is characterised by the dismantling of Medibank at the hands of the Liberal–National Country Party Governments. Table 6.1 presents a summary of the changes which have occurred between 1953 and 1 September 1981. In that time Australia's health care system has swung from being almost entirely dependent upon private expenditures to one heavily dependent upon public (Medibank) expenditure and thence back to arrangements very similar to those obtaining before the advent of Medibank.

As far as health care provision is concerned the era 1950 to 1969 is dominated by the arrangements primarily sponsored by Sir Earle Page, the central phase of which was the Voluntary Health Insurance Plan introduced in 1953 which entitled Australians to government subsidy of health care costs provided they were voluntary members of private health insurance funds. Similar arrangements had been made for insurance against hospital costs in 1951. Page (1960, 40–1) describes the general objectives of his plans thus:

To ensure everyone in Australia who makes provision through insurance will be able to cover the major cost of all surgical and medical attention, whenever or wherever sickness or accidents occur . . . to qualify for benefits you must be a financial member of an approved medical insurance organisation so long as you pay your small weekly premiums, then you and your dependants will be entitled to the benefits. The Commonwealth Government will provide financial support of the voluntary insurance against its costs of medical attention. This means in effect that if you insure yourself the Commonwealth will provide generous financial help. But the Medical Benefits scheme is not a plan for compulsory health insurance. It is deliberately designed to strengthen voluntary medical insurance by helping those who wish to help themselves.

Herein is the clearest annunciation of the belief that health care provision is based on an individualistic conception of man, the state's role being confined to that of a supporter of individual and private effort. This policy ensures the principal objective of rationalising the supply of services at the minimum cost to the state. So successful were these arrangements it was nearly 20 years before they were seriously questioned. Scotten (1974) reports that his own 1968 examination of the 'Page Scheme', as these arrangements generally became known, found that operational efficiency and welfare optimality were not being achieved. In other words the benefits of the scheme were no longer sufficient in the face of growing expectations amongst the Australian population which wanted a simpler and more comprehensive system of health care. The cases for such reforms was made in 1969 by the establishment of a Committee of Inquiry under the Chairmanship of Mr Justice Nimmo. The report of the Nimmo Committee was very influential in highlighting the deficiences of the Page Scheme, which were as follows:

1. The scheme had become too complex and was incomprehensible to many people.
2. Benefits paid to contributors were often less than the cost of treatment.
3. The cost of private insurance contributions had grown to a point where they were too expensive for many Australians.
4. The private insurance schemes often excluded patients with chronic conditions, and pre-existing conditions, and required them to pay additional premiums, both of which caused hardship.

5. Administration often absorbed too great a proportion of all contributions.
6. The level of reserves held by insurance funds was too high.
7. Services ancilliary to medical and hospital care, for example dentists, chiropractors, etc., were either excluded or required additional payments.

The Nimmo Report made 42 recommendations for reform, out of which a New Health Benefits Plan emerged in 1970, sponsored by the then Liberal Government. Under this new scheme – which still relied heavily on private health insurance funds – the benefits paid to members were increased so that no more than $5 was paid out of the patient's own pocket. This was achieved by the government agreeing with the Australian Medical Association a scale of fees to be charged. Provision for chronic illness and pre-existing illness was extended and low income groups could apply to have their health insurance costs met by the state. The initial cost to the state of these reforms was estimated at $16m but the final total was of the order of $100m (Hetzel, 1980). These amendments were minor, however, compared with the major change in the role of the state as provider of medical care when the Medibank scheme was created in July 1975.

Main functions of the scheme were universal entitlement to benefits without any means test including basic medical and hospital care. Finance was administered by a separate statutory Health Insurance Commission with funds derived from general taxation. In practice this meant treatment in public hospitals became free, subsidies were paid to patients in private hospitals and to private patients in public hospitals. In general practice patients received treatment from doctors who billed the Health Insurance Commission and were paid on a fee-for-item of service basis according to a scale of agreed charges. Payments by patients were not to exceed $5 per item. The cost of the hospital aspects of the medical scheme amounted to about $800m in 1975/76 (Hetzel, 1980). In addition the Labour Government also rigorously pressed on with the development of a Community Health Programme which advocated prevention and rehabilitation schemes (Australia, 1973).

This period represented the zenith of state involvement in Australian Medical Care; it was not an involvement easily won and there remained a great deal of opposition. In attempting to pass the

Medibank legislation the Labour Government was forced to dissolve both the House of Representatives and the Senate of the Parliament and seek re-election in May 1974. Although it won, the Labour Party still lacked a majority in the Senate where its earlier attemps to enact Medibank had foundered. But because the Bills concerned with Medibank had been the cause of a double dissolution and because they had been rejected twice by the Senate (after an interval of three months) both Houses of Parliament held (as they can under the Australian Constitution) a joint sitting for the purpose of passing the rejected Bills. It was only after this process (used for the first time ever) that Medibank legislation became law (*The Lancet*, 1975). Even so, not all Australian States (those whose own State Parliaments were controlled by parties other than Labour) were prepared to co-operate with the Federal Government and make the scheme fully operational. For instance, 'The situation in Queensland would be farcical if it was not so tragic. The County (Farmers) Party, which is the Senior coalition partner with the Liberals [in Queensland] (Right-of-Thatcher Conservatives) is adamant that it does not want "socialist centralist money"' (*The Lancet*, 1975, 1332).

Within a year of the publication of this viewpoint the process of dismantling Medibank began. In December 1975 a Liberal Government returned to power and, despite promising to leave Medibank intact, a number of changes to Medibank were introduced on 1 October 1975. Medibank Mark II, as the new scheme became known, was not to be financed from general taxation, but from a 2.5 per cent levy on income, though Australians could opt out of paying this provided they joined a private insurance fund. Medibank itself became a government-sponsored health fund in competition with other private funds, adopting the appropriate name of Medibank Private. Since then the lessening of state involvement has continued. In July 1978 Medical benefits were reduced from 85 per cent of scheduled fees to 75 per cent and in November of the same year health insurance ceased to be a compulsory requirement. Instead a Commonwealth Medical benefit equivalent to 40 per cent of approved fees was payable to all patients (Medibank Mark III). One year later this scheme was replaced; in the new scheme all doctors' charges over $20 were refunded, and public hospital charges increased. In July 1981 contributions to health funds became eligible for tax relief of 32 per cent on premiums paid, and then on 1 September 1981 the

Commonwealth Government transferred responsibility for health services back to the states with the intention of running down Federal funding of the hospital services. Free treatment in public hospitals was abolished except for those people defined as 'disadvantaged' – including pensioners, the unemployed and recent migrants – who were to apply for a medical card exempting them from charges. Approximately 20 per cent of the population fall into these categories – the remainder had to decide if they were prepared to risk paying 100 per cent of charges or take out insurance to cover them, basic cover costing about $550 per annum (O'Sullivan, 1981, 8). Announcing these changes in the Australian Parliament the Commonwealth Minister of Health Mr McKellar proclaimed them as necessary because, 'those people in the community who can pay their own way should be required to contribute to the costs of the health services they receive – either by personal payment or through health insurance' (quoted in *The Advertiser*, 30 April 1981, 1). A comparison of this 'user-pays' perspective with Sir Earle Page's description of his own medical care scheme demonstrates how the financing of health care in Australia has come full circle. Consumers have, not surprisingly, been baffled by the changes which have occurred (with six modifications to the Medibank scheme between October 1976 and September 1981) and some observers have asserted it is they who will suffer most (Bates, 1980).

How can we account for these changes? Why has Australia swung so violently from private to public and back to private provision? Hetzel (1980) advances two reasons given by the Liberal Government when it first began dismantling Medibank. It was argued that Medibank had become too costly because as a state-run bureaucracy it was inefficient, and because providing state subsidies encouraged over- and mis-use of medical care. Hetzel dismisses these arguments. In his view the nature of health care prevents true competition. Not only do the major providers of care – doctors – not engage in price competition, but the consumer with imperfect knowledge is faced with considerable problems in judging the quality of service available. Moreover, Hetzel argues actual Medibank experience showed health administration costs to be only about 30 per cent of the previous privately based system. Turning to the question of over-use by Medibank patients, Hetzel (1980, 277) quotes Deeble, who concluded, 'The extension of hospital and medical insurance coverage to the whole population had little effect on utilization overall, but may have resulted in some redistribution of services towards previously disadvantaged groups.'

Two alternative reasons are advanced by Hetzel. First, the Liberal Party was subjected to pressure from traditional supporters with interests in the health field – doctors who feared the state as a threat to their clinical autonomy – and the private health funds. Dismantling Medibank was obviously in the interests of these groups. Secondly, one of the major objections of the Australian Commonwealth Government in dismantling Medibank has been the reduction of public expenditure. Thus amendments to the provision and financing of health care can be seen as one dimension of wider public expenditure policies which seek to redefine the relationship of state and private expenditures in the belief that Australians should be responsible for their own health and health care.

Indeed, in the Australian example, we can use the two political parties to illustrate how conceptions of man and society, serving specific economic, political and institutional interests, can shape the nature of health care provision. The party political differences are only one part of the process because the changes in provision they enact represent different cultural definitions of man, health and society and different conflicting power alignments.

Health Care in Britain: A Case of Unfulfilled Promises?

Every so often Britain's social conscience is pricked by reports of ill-treated patients, mismanaged cases which result in the injury or death of young children at the hands of their parents and appalling conditions in long stay hospitals. The reaction is real enough and there is undoubtedly a sense of collective guilt that things happen to patients in hospitals and other institutions managed by the NHS. But the frequency with which these reports appear and the number of enquiries which continue to reach the similar conclusion about overworked staff and inadequate facilities suggest the sense of guilt soon passes and that our concerns are never quite converted into action.

The 'Cinderella Services' for the mentally ill, mentally handicapped and the elderly are the victims. Despite repeated demands to direct a greater proportion of NHS resources towards these groups in order to promote better standards of care the reality is one of failure. The problems might not be so intense if expenditures on the NHS were growing rapidly, but they are not. Any policy which seeks to support the Cinderella Services must therefore spell out from where within the NHS it will derive its resources. Dr David Owen, then Minister of State for the NHS, put the point succinctly:

> We must cease in the health service demanding more for one sector (say for renal dialysis) without recognising that if we take more from one sector it has to come from another . . . we must be prepared to say, if we want priority for one sector, where the money should come from. (Owen, 1976, 113)

In 1976 Dr Owen's department went so far as to quantify the shift of resources from one sector of care to another (see below) but as with other exhortations to spend more on the psychiatrically ill, more on the disabled, more on community care, but less on the acutely sick and less on hospital care, 'some of the changes recommended in the plan are simply not materialising, or are materialising so slowly as to be difficult to discern' (DHSS, 1980, 236).

Before looking at some of these plans in detail there is an important distinction which must be made. When talking of the distribution of resources in the NHS we can talk either of their allocation to particular 'sectors' or 'programmes'. The former are groupings of particular types of service, for example hospital, and may include, amongst others, acute, maternity and mental illness hospital services as distinct from community services including health visitors, home helps, etc. The latter are groupings of patients, for example the elderly, the mentally ill, etc., who consume services drawn from each of the health care sectors. It is in a sense an accounting difference but it enables us to distinguish expenditure on people from those on services. In this presentation we discuss expenditure by sectors and programmes together because the general comment that progress in changing them has been slow is equally applicable. The care of the mentally handicapped and the mentally ill are illustrative.

Two White Papers of significance to the care of the mentally handicapped and the mentally ill were published by the DHSS during the 1970s (Gt Britain, 1971, 1975). The first of these recommended a major transfer of pa ients from long stay hospitals (with in total over 50,000 beds) to community care. Over a 20 year period the idea was to reduce the number of adult in-patients from 51,000 to 22,000, whilst at the same time increasing the number of residential places in the community from just over 5,000 to over 29,000. It is reported, however, that on the basis of the rate of change achieved between 1971 and 1976 it would take 29 more years to reduce the in-patient load and 17 more to increase the community care capacity to target (Tyne, 1978).

The second set of recommendations were made with regard to services for the mentally ill. The 1959 Mental Health Act advocated that local authorities should begin to provide day care and residential services for this group of patients and this set in motion a decline of the psychiatric inpatients population of 20 per cent between 1967 and 1973 (Gibbons, 1978). But, whilst this indicates a shift away from hospital care, Gibbons argues that alternative provision in the community, as advocated by the 1975 White Paper, has not materialised, with the result that some ex-patients end up sleeping on the streets, or in lodging houses and in prisons. Likewise families probably absorb a great deal of the social cost of this failure. As a result many new psychiatric patients will still end up in old psychiatric hospitals isolated from their homes. As Gibbons (1978, 121) concludes, 'the reality is that old mental hospitals and institutional psychiatry will continue to have the major responsibility for the care of the more severely and chronically ill patients for a long time to come'.

The publication of 'Priorities for Health and Personal Social Services in England' (Gt Britain, 1976a) re-emphasised the objectives of the 1971 and 1975 White Papers, but it went further in specifying the growth in current and capital expenditure which would be made available to the NHS and local authorities to implement the programme. At the same time it also indicated how this extra funding would be raised. The consultative document indicated that between 1975/6 and 1979/80 current expenditure on health and personal social services would grow by just over 2 per cent, but high priority programmes would receive rather more than this, at the expense of low priority programmes. Thus services used mainly by the elderly were to receive an extra 3.2 per cent per year, the mentally ill 1.8 per cent, the mentally handicapped 2.0 per cent and children's services 2.2 per cent per year. In contrast acute and general hospital services would only receive an extra 1.2 per cent per year and hospital maternity services would actually lose funding at the rate of 1.8 per cent per year. The great majority of the extra expenditure was to be devoted to the development of primary and community care sectors, and the document spelt out in some detail the number of new residential places, day care centres, training services and so on to be provided annually. Figure 6.1 presents the distribution of health and personal social service capital and current expenditure by each of the programmes described above as a proportion of all expenditure. It will be seen that the DHSS was predicting that expenditure on services for the mentally handicapped would rise from the 1970 figure of 4.4 per cent to 4.9 per cent in 1975/6 and thence to 5.6 per cent in 1979/80. Correspondingly, expenditure on general and acute hospital and maternity services would fall from 46.0 per cent in 1970 to 43.1 per cent in 1975/6 and 40.7 per cent in 1979/80. Specification of precise allocations suggests some rational basis for their selection, as though underlying the identification of a figure of 5.6 per cent for mental handicap was some objective evidence of quantified needs. However, they are:

> not the result of the accumulation of knowledge of the best use of national resources in securing a healthy society, but more the result of the historical interplay of the health care professions, central and local administration and party and public opinion. (DHSS, 1980, 240)

This interplay was clearly demonstrated by the comment that 'in times of economic restraint plans for the changes outlined must be held over until funds can flow more freely, so as to avoid depriving the barely adequate in favour of neglected areas of care' (BMA, 1978, 254). Such perspectives were not limited to the BMA though, Gt Britain (1977) reported that some community health councils were anxious that services for the economically active population should not suffer in spite of the priority groups claims. Reviews of the priorities policy agreed with the general direction adopted but the Black Report was unhappy about the progress which had been made towards the achievement of priority objectives. Table 6.2 shows that personal social service expenditure on community care actually declined from 22.7 per cent of all expenditure to 21.1 per cent between 1974/5 and 1977/8. Residential care expenditure (which should decline) actually increased from 49.6 per cent to 50.6 per cent.

Figure 6.1:   Health and Personal Social Services Expenditure (Capital and Current) by Programme (£m November 1974 Prices)

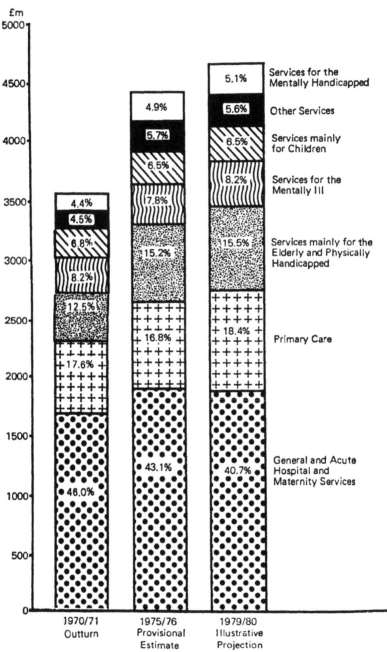

Source: Gt Britain (1976a).

Table 6.2: Distribution of Expenditure Within the Personal Social Services

Year	% Distribution of local authority expenditure				
	Residential care	Community care	Day care	Other	Total
1974/5	49.6	22.7	11.3	16.5	100
1975/6	51.0	21.7	11.4	16.0	100
1976/7	50.8	21.0	11.8	16.4	100
1977/8	50.6	21.1	12.0	16.3	100

Source: DHSS (1980), p. 235.

Citing evidence from a subsequent public expenditure White Paper, the Black Report shows that the publication of the priorities document in 1976 was significant in other respects. Having spelt out the need for greater expenditure on personal social services in 1976, that year ironically marked the end of growth in expenditure on these services. In 1973/4 expenditure grew by 15 per cent; in 1974/5 and 1975/6 by 8.3 per cent and 6.5 per cent respectively; but in 1976/7 and 1977/8 it actually declined by 1.3 and 1.6 per cent respectively, following the Labour Government's decision to cut back on public expenditure in the face of the fiscal crisis precipitated by world reaction to large-scale public sector borrowing. The inescapable conclusion to which this analysis leads is that the establishment of priorities does not guarantee their achievement. Specifying them does, however, have advantages; it quite openly admits to perceived deficiences and creates a yardstick by which progress towards their achievement can be met. But this can be embarrassing; no government enjoys having its own objectives quoted back in order to illustrate their failings. Perhaps this is why the latest statement from the DHSS has moved away from the previous approach of precise specification. In Gt Britain (1981) the current Conservative Government reaffirms the elderly, the mentally ill, the mentally handicapped and children as priority groups. Expenditure on community care is also advocated but the document avoids the specification of particular levels of service growth as objectives. Rather the approach is much more nebulous; it restates the characteristics of a desirable service pattern – locally based, community orientated, etc. – but does not specify rates of change in favour of particular programmes or sectors. What is clear, however, is that progress in the direction of the 1976 and 1977 documents is likely to be even slower. This must be the case given government public expenditure plans for the personal social services (Table 6.3). Although total health and personal social services expenditure was predicted to grow by 3.6 per cent from 1978/9 to 1981/2 the proportion to be spent on personal social services actually declined from 16.1 per cent (1978/9) to 14.9 per cent (1981/2). The extra 3.6 per cent in total expenditure between 1978/9 and 1981/2 finds its way predominantly into hospital and community services, but this is unlikely to be available for switching the direction of NHS spending because it is generally reckoned to cost an extra 1 per cent per year to accommodate the cost of an

Table 6.3: Health and Personal Social Services Gross Expenditure (England £m at 1980 Survey Prices)

	1978/9	1979/80	Provisional outturn 1980/1	Planned 1981/2
Health				
Hospital and community health:				
current	5,405 (57.9)[a]	5,407 (57.7)	5,510 (58.2)	5,635 (58.3)
capital	434 (4.6)	395 (4.21)	468 (4.4)	454 (4.7)
Family practitioner services	1,698 (18.2)	1,704 (18.2)	1,744 (18.4)	1,789 (18.5)
Central health services	287 (3.07)	304 (3.2)	339 (3.6)	341 (3.5)
Total health	7,824 (83.9)	7,810 (83.4)	8,011 (85.5)	8,219 (85.0)
Personal social services[b]				
Local authority:				
current	1,438 (15.4)	1,486 (15.8)	1,370 (14.5)	1.364[c] (14.1)
capital	57 (0.6)	59 (0.6)	70 (0.7)	69 (0.7)
Central government	7 (0.7)	8 (0.8)	9 (0.9)	8 (0.9)
Total PSS	1,502 (16.1)	1,553 (16.5)	1,450 (15.3)	1,442 (14.9)
Total HPSS	9,326 (100)	9,364 (100)	9,460 (100)	9,662 (100)

Notes:
a. Figures in brackets are percentage of column totals.
b. Figures for 1980/1 and 1981/2 tentative as they depend on local authority decisions.
c. Expected to be greater than 1980/1 by transfer for the Urban Programme.

Source: Gt Britain (1981), p. 5.

ageing population and medical advances. In other words, the 3.6 per cent is intended to enable the NHS to stand still, not change direction. In short, the priorities policy has effectively been abandoned, or more accurately the new priorities are the perennial ones – those of acute and hospital medicine.

The key question again is why has the policy failed? It is tempting to believe that the reason is lack of will to change existing patterns of provision or that the administrative mechanisms of change are inadequate to make the policy effective. Heywood and Alaszewski (1980) have argued from this perspective, asserting that the failure to achieve central government strategies is a consequence of the failure to recognise the importance of local medical hegemony in the workings of the NHS where:

> the agenda, especially at local level, has been controlled by the medical profession. Thus, local politics are medical politics. The upper reaches of the NHS may be more open to new policy initiatives, but they are rapidly filtered. By the time they reach the local agency, they have been readjusted to fit the local political agenda. This can be seen most clearly in the planning system which was an attempt by the centre, to change the way the political agenda was formed. The guidance made it clear, for example, that consideration should be extended to existing commitments as possible sources of monies for needed developments. The centre also tried to use the planning system to put the case for the cinderella services firmly on the local agendas. Our descriptions of these efforts suggest that they can hardly be described as a resounding success. The proper designation is really 'failed', particularly since public expenditure plans for 1980/81 envisage an absolute decline in spending by Social Services Departments. They are responsible for developing community services which are disproportionately used by the Cinderella groups, and the cutback in resources indicates an abandonment of the effective priority for them. It is another recognition of the centre's inability to have an effective influence on the local agenda. (Haywood and Alaszewski, 1980, 136–7)

Adopting this perspective suggests the policy could be implemented if the centre recognised the importance of medical power at the local level and was prepared to do something about it.

We agree with Haywood and Alaszewski that medical power is certainly a dominant force, especially at a local level, but we are not convinced it is a wholly satisfactory explanation for the failure of centrally formulated policy priorities. Indeed, Haywood and Alaszewski hint at a much more convincing explanation when they comment on the decline of funding for social services to which we have referred above. Agreed, medical hegemony protects established patterns of service when the choice is between cutting them for the sake of, say, mental handicap, but this suggests the funding of the NHS is somehow separated from the sociopolitical context in which it is located. A much more fundamental question is why the state felt obliged to cut back on social service spending and in so doing renege on the priorities policy. It suggests that certain interpretations of social policy needs may have little impact on actual expenditure because they do not command the support of important interests – political, professional or economic. Further, as in our Australian example, the question of the state's interests and objectives as provider of health care is raised.

Health Planning and the Role of the Capitalist State

The capitalist state is not an autonomous entity. As well as the nature of administrative practice, the nature of the state and its relations to other societal institutions, predominantly the economy, limit the potential for reform from within. No government can ignore or evade these systemic constraints for long, even if it wants to. These constraints are determined by the nature and requirements of the mode of production. Thus a capitalist economy has its own imperatives to which any governments and state must sooner or later submit (see Poulantzas, 1973). Poggi (1978), in fact, notes the identification of the modern industrial state with economic management and with economic interests. But by merely providing the conditions for the effective abstraction of labour (appropriating surplus value from the workers) from the means of production, the state represents the apparent separation of the political from the economic under capitalism. The state acts, in Poulantzas' terms, as the factor of cohesion within a social formation. Indeed its success depends upon it ensuring the cohesion of that formation (the social, political and ideological superstructure) and the reproduction of the capitalist mode of production. This superstructural role is vitally important with respect to many intermediate social groupings – the middle classes, administrators and bureaucrats (e.g. within the NHS) – that occupy an intermediate position within the social relations of production. They occupy a contradictory class location, serving the interests of the ruling class (see Wright, 1978). Such groupings have an important legitimising function in their own right with their role being to make decisions, arbitrate, allocate resources (on behalf of the state) which thus appears as an independent phenomenon.

It is possible to argue that the functions of the state demand its relative autonomy in order to organise the interests of the dominant class and frustrate those of the subordinate class, thereby sustaining the process of capital accumulation against its contradictions.

> The contradiction determining the state's centrality to capital accumulation in the monopoly phase is the falling tendency of the rate of profit. This entails the constant restructuring of the capital relation through increased state expenditure. This in turn causes the contradictions to capitalism to be assimilated to the state apparatus, which increasingly reflects the basic capital relation and . . . appears increasingly disinterested. (Cooke, 1980, 212)

State expenditure, therefore, constitutes a necessary investment for reproducing the conditions of accumulation (see Hirsch, 1978). Indeed O'Connor (1973, 6) identified two, often contradictory, functions of the state – accumulation and legitimisation: 'The state must try to maintain or create the conditions in which profitable capital accumulation is possible. However, the state must also try to maintain or create the conditions for social harmony.' This analysis is taken further by Habermas (1976) who suggests that organised and state-regulated capitalism is marked by the state intervening in the market as functional gaps appear. The state carries out numerous imperatives of the economic system; it not only plans and regulates the economic cycle as a whole, but through administrative procedures attempts to avoid instabilities. The state also 'actually replaces the market mechanism whenever it creates and improves conditions for the realisation of capital' (Habermas, 1976, 35) through, for example, improving the material infrastructure (education, health, transport programmes) and relieving the social and material costs resulting from private production (welfare payments, environmental protection). As the state now reflects the capital relation, or in Habermas's terms as the relations of production has been repoliticised, there is an increased need for legitimisation – a problem solved by a system of formal democracy in which resources are allocated by rational-legal decisions on equity criteria. The state thus becomes regarded as the representative of the common interests of the people. Thus, as Gough (1979) argues, the welfare state has an ideological role. Its social expenses – the maintenance of the non-working population – are not even indirectly productive for capital but are a necessary expense, helping to fulfil the state's legitimisation function and, therefore, helping to maintain harmony. Other expenditures – social investments (increasing the productivity of labour) and social consumption (lowering the reproduction costs of labour power) – augment the process of accumulation. Thus, 'the welfare state denotes state intervention in the process of reproducing labour power and maintaining the non-working population' (Gough, 1979, 49). The provision of medical care is, of course, an important part of this process, but as Offe (1972) argues certain groups and vital 'areas' are excluded from enjoying the services of public bureaucracies:

> This seems to be symptomatic of a phase in capitalist development in which areas of crisis peripheral to the central groups of problems [economic stability, foreign and military policy and mass loyalty] . . . are hindered from generating further disturbances to the system, but are otherwise left to themselves.

There appears, therefore, the modern pauperism of depressed 'areas' – education, transportation, housing and health – which effects the entire population. Marginal institutions and groups – the unemployed, the pre-school population, the retired and the mentally ill – also belong to this category.

Navarro (1976) sees the growing socialisation of production requiring continued state expenditure to facilitate private capital accumulation, but in so doing generating a fiscal crisis in which the additional expenditure cannot be met from increased revenue. This:

> continuous fiscal crisis of the state results in: (a) a cutting of social wages; (b) an increasing demand for planning of the economy, which requires still further centralisation of power in

contemporary capitalism; and (c) a growing demand for rationalisation of the system, which calls for higher productivity and efficiency. (Navarro, 1976, 221)

In the medical care sector this means cutting the rates of growth in medical care expenditures – either by transferring responsibility back to the individual and the private sector, or by abandoning planned expansion of services which are funded in Gough's (1979) terms as social expenses. The argument advanced to support these strategies emphasises the essentially unproductive nature of public expenditure and the need to shift capital into the private productive sector as a means of ensuring further capital accumulation.

Thus, by examining the role of the capitalist state, we can discern certain systemic imperatives that appear to limit the nature and amount of health expenditure and provision whatever the conceptions and definitions of health. The limited reforms in Australia and the failure of priorities policy in Britain suggest that power relations greatly shape the nature of care. Altered power relations may change the nature and distribution of care. Thus systemic imperatives are not eternal nor are ideas and conceptions unimportant. Such imperatives severely limit the scope of change. But ideas coupled with the experience of the disadvantaged may well dislodge an embedded system of power and privilege (see Chapter 7). Such dislodging occurred in our next example, the Soviet Union. In this case, however, it is possible to suggest that there developed new imperatives of the 'socialist' economy and polity; the reproduction of the centralised system of production and of Soviet political relations.

Soviet Union

The inclusion of the Soviet Union in the category of advanced industrial society may appear to be problematic. Indeed, it can be argued that it has a great deal in common with the other state socialist societies considered in the next section. Unlike the capitalist societies, already discussed, with their origins in the nineteenth century, state socialism is a twentieth-century phenomenon. It came to societies which had reached only low levels of economic development, and where the peasantry made up the bulk of the population. State socialism 'refers to any economic order in which the means of production is formally socialised in the hands of the state' (Giddens, 1973, 155). It provides an alternative framework for channelling the industrialisation process. In many ways the Soviet Union, like the USA is the most advanced example of the type – an advanced industrial nation of the state socialist variety. Indeed, the inclusion of the Soviet Union in this category points to one of the major contentions of this chapter, that the social developments, including those in the health care field, are not to be understood in terms of divergent cultural values and definitions alone but also in terms of persistent differences in social, economic and political structure, that is in terms of societal constraints.

The Soviet health care system was constructed from a low and differentiated level of provision that reflected the largely feudal conditions of pre-revolutionary Russia. The rural population, in particular, received scant medical attention. In 1913, there was a ratio of 15 doctors per 100,000 people compared with a 1910 ratio of 157 for the USA (Field, 1967). Thirty-five per cent of towns had no hospital. The Soviet state considered the improvement of health care provision to be vital (see Field, 1976; Smith 1979), not merely to equalise such provision but to assist in the development process. A healthy workforce was seen as a necessary prerequisite for increasing industrial production and many facilities were and are provided at the workplace. Health is thus seen to have a societal function. It is conceived of in terms of social rights and of systemic improvement (see Chapter 2). Such objectives mean that there is not perfect equality in distribution. Smith (1979) points out that as long as medical services are subject to a degree of scarcity, those who perform the most important jobs are entitled to priority treatment. We should add that the relative importance of occupations is usually determined by those in dominant positions. Thus Smith (1976) points to the special facilities enjoyed by the political and intellectual elites in Moscow and Ryan (1978, 111–12) argues that 'the Soviet state has institutionalised for an elite high-quality provision which by its very nature is rigidly exclusive and can only serve to perpetuate, if not harden, existing socioeconomic differentials'. At the other end of the scale of provision, in terms of quality and quantity, are the facilities provided for farm workers (see Kaser, 1976).

Indeed, the Soviet system has concentrated on equalising provision between town and country and between the constituent republics. Hyde (1974), in fact, suggests that local variations in health conditions and needs may require departure from the application of centrally determined norms for resource allocation, as for example in the development of Siberia (see Chapter 3). It can be argued though that the central planning and allocation of health care through these norms – the approved number of beds, doctors, clinics, etc. per 1,000 population – has helped to improve, maintain and equalise provision. Thus in 1913 the ratio of the highest to lowest republic rates for the provision of physicians per 100,000 was 45.00 and 65.50 for hospital beds. These ratios fell to 3.76 and 2.25 respectively in 1940 and 2.00 and 1.43 in 1975 (Cole and Harrison, 1978; Smith 1979). But both Kaser (1976) and Zwick (1976) point out that it has been easier to provide health care facilities rather than personnel *in situ* as the latter dislike living in developing or agricultural areas. Soviet doctors seem, therefore, to have residential preferences, like those of American doctors, that affect the provision of health care. Professional constraints, therefore, operate in a centrally planned as well as a private enterprise system. Indeed Ryan (1978) calculated that in the Soviet Union there were 345 doctors per 100,000 urban population and 179 per 100,000 rural population. It is interesting to examine such differences with the convergence of republic ratios of hospital beds to population (see Figure 6.2).

The differences in facility and professional provision are also identified by Cole and Harrison (1978) in their examination of physician provision in 1975 and hospital bed provision in 1974. For physicians, Georgia, the best served republic (411 per 100,000), has about twice the ratio of the worst off Taszhikistan (206), though its ratio is worse than that of the best regions – Central (459) and North West (455) in the Russian Republic. The range for hospital beds is less – 87 in Armenia to 124 in Latvia. Again the regions of the Russian Republic perform remarkably well not only around Moscow and Leningrad but also east of the Urals, again supporting the points that professional constraints operate and that it is easier to build facilities than staff them in particular areas, even in centrally planned systems.

The central planning of the Soviet system takes the form of a hierarchy of facilities and services. Smith (1979) describes how at the base of the system, urban and rural areas are divided into microdistricts (*uchastok*) of about 4,000 inhabitants. Such districts are served by two physicians, a paediatrician and one or two nurses. Ten of these districts (*raion*) typically combine to provide primary care at a centralised outpatients' clinic. Hospital provision depends on population density. The basic unit of the Soviet system is, in fact,

Figure 6.2: The Convergence of Republic Ratios of Hospital Beds to Population, Expressed as Percentages of USSR Ratio

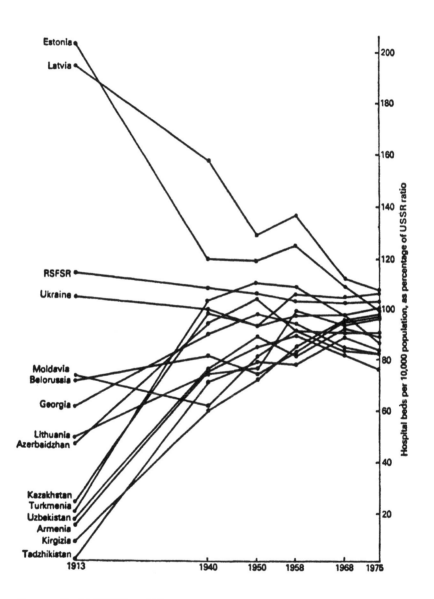

Source: Smith (1979), p. 297.

the 'polyclinic' generally attached to the (district) hospital, which offers comprehensive outpatient and health centre services. Flexibility is given to the health care system by:

> the emergency medical care stations, from which ambulance services and mobile personnel operate; the feldshers or doctors' assistants, who provide services short of that of a fully trained physician; and mobile (e.g. airborne) services for more remote rural areas. It should be noted that many Soviet Citizens have access to medical care at their place of work, both in factories and on farms. (Smith, 1979, 293)

Such norm-based health care provides rough equivalence of provision, predicated as the system is on the overt egalitarianism of marxism-leninism as a practised political philosophy, on the need to maintain and improve human capacity for production purposes and on the importance of health for the evolution of socialist man. The socialist state has its own imperatives rather like those of the capitalist one. Referring both to tax and social security systems and to health and education, Titmuss (1974, 17) argues that:

> Soviet Russia . . . has fashioned a model of social welfare which is based, in large measure, on the principles of work-performance, achievement and meritocratic selection . . . Soviet social services – like the mental health service – are, in part, social control mechanisms in respect of dissenters, non-conformers, deviants and underachievers. In part, they also function to sustain and glorify the work ethic . . .

Work and welfare are as inextricably connected in Soviet society as they are under capitalism (see also Madison, 1968; Doyal (1979b) assesses the relationships between workplace and health under capitalism). The socialist state is more than an adjunct to the economic system. It is that system. The state directs, plans and controls but in so doing aims at the social integration of the population. Its ideology emphasises worker and community provision of health care. Such an emphasis demonstrates that, like the capitalist state, the socialist one is based in part on the past. Thus its advocacy of prevention is based on the pre-revolutionary practice of serving the countryside by medical auxiliaries. These feldshers have direct parallels in other state socialist societies, like China and Cuba.

Resource Limitations, Rural Development and Preventive Medicine

We now turn to a consideration of the problems of providing health care facilities in developing societies. We have already, in passing, mentioned such problems in considering the change from traditional to modern systems of care in such nation-states as Ghana and the Ivory Coast. We wish now to concentrate on societies that have rejected the necessity for the predominance of a Western-style system and that have been influenced by the Soviet model, especially its emphases on rural provision and the preventative role of the feldshers. Western or scientific curative medicine is seen as expensive and elitist in that provision is limited to a few localities and is likely to benefit privileged minorities. Such a concentration of health facilities is regarded as immoral when the health status of the mass of the people requires much improvement.

The post-1949 Chinese state began with a population assailed by many infectious diseases, a life expectancy at birth of 30 years, infant mortality at a rate of 200 per 1,000 and about 40 million opium users (Manzie, 1979). In 30 years Chinese health policies have achieved the mortality pattern which took the West three times as long to obtain. Indeed, the leading killers in China now are cardio-vascular disease, lung cancer and accidents – the so-called diseases of affluence. To achieve such results 'the Chinese have employed principles particularly suited to the historical, cultural, geographical, and political conditions of the society' (Sidel, 1977, 194). This has meant an explicit rejection for a health care system based predominantly on the curative-technological medical model of the West (see Wilenski, 1977). Thus the central features of the Chinese health system are as follows:

1. Medical services are linked with production and there is an attempt to dovetail them with the economic structure. The work ethic and the will to establish a socialist society are therefore central to understanding the Chinese system. Improvements in health care provision in the countryside are centred on the rural production team of 50 to 200 people which is served by one or two health aides and barefoot doctors, and the production brigade of 1,000 to 2,000 individuals with two barefoot doctors. Such teams and brigades are also served by commune hospitals, provided by a combination of rural self-help and direct assistance – mainly in

the provision of Western-style doctors – from the urban areas.

2. Priority is given to prevention by mass methods (see Byers and Nolan, 1976). The emphasis on prevention started in 1949 in the tackling of infectious diseases with a reliance on mass education rather than professional teams. Investment has been directed towards improved sanitation methods rather than prestigious medical equipment. Thus, for example, the standard fertiliser (human excrement) is now composted to destroy pathogens and new canals have been dug to replace those infested with schistosoma-carrying snails.

3. New kinds of health professionals have been employed. The barefoot doctors and medical aides are hygienists and health educators who give care on a part-time basis. Little is known about their clinical prowess but their manual includes basic anatomy, some acupuncture and massage techniques, descriptions of basic diseases, lists of traditional remedies and reference to herbal botany. They are concerned with prevention in the main. They also act as 'gatekeepers' to the centrally provided curative facilities, enabling the limited number of doctors to concentrate on those who most need their services.

4. There is an integration of traditional and modern medicine (see also Sidel and Sidel, 1973). In 1949, the Chinese state inherited at most 40,000 Western-style doctors and about 500,000 traditional healers. In 1965, a directive from Mao initiated not only the development of rural health care but also promoted Chinese medicine which dates from around 800 BC. In fact it is now possible to find moxibustion, cupping and the intravenous injection of herbs in district hospitals. Smith (1979) rightly points to the preoccupation of Western interest in Chinese medicine with the technical curiosity of acupuncture – a preoccupation which says more about the Western conceptions of health and health care than those of the Chinese. Increased emphasis on Chinese medicine means not only cheaper provision but also the deprofessionalisation and demystication of medical practice and mass participation in the provision of health care. Health care is again an adjunct of the interests of the social order, predicated as such interests are on specific models of man and society.

5. There is a policy of selecting mature students for university medical training. Until recently, candidates had to serve for

up to five years in agriculture, in industry or the army. This policy may well have changed under the current, more market and specialist-orientated Chinese leadership.

Many of these threads of policy are brought out in Lee's (1981) study of medical care in Toushan commune, Kwangtung province. This mainly agricultural commune had in 1978 58,000 people living in 137 villages and organised in 21 production brigades and 187 production teams. There are in fact three levels of health care provision – commune, brigade and team. At the team level, there are simply auxiliary aides recruited from the peasantry. They are mainly engaged in agricultural work but are part-time assistants to the barefoot doctors and are responsible for first aid, immunisation, environmental sanitation and nursing care in their own teams. At the brigade level, there is a medical station, encompassing about 350 auxiliary workers. Each brigade also has three to five barefoot doctors, responsible for preventive medicine and the treatment of common and relatively minor illnesses. There are no inpatient facilities at this level. At the commune level, there is a health clinic, the locus of physician treatment, as well as a medical station for the industrial and transport workers. This station is responsible for the prevention of occupation diseases and outpatient care of the industrial workers. It is staffed by five barefoot doctors, making a ratio of provision of 1 to 460 people. This is compared with 1 to 586 in the agricultural areas. There is a total of 25 doctors attached to the health clinic. This figure includes both Western and Chinese-style and professional and assistant physicians. The doctor to population ratio is 1 to 2,317. There are fewer Chinese-style than Western-style doctors. This is reflected in the dispensing of medications (70 per cent Western) and the choice of patients at the clinic's outpatient departments (70 per cent Western). But 80 per cent of *all* outpatients visits are to the Chinese style barefoot doctors. Indeed, while at brigade level Chinese treatment is emphasised, there is a division of responsibility at clinic level. Western-style doctors concentrate on surgery, dermatology, paediatrics, gynaecology and immunisation, while Chinese specialise in acupuncture and moxibustion and the treatment of sprains and contusions. It is possible to argue, however, that Western medicine is the basis of health clinic practice. There is though a flow of patients, knowledge, personnel and technology between medicines and between levels.

In its entirety, the Chinese system can be said to stress prevention and the provision of low-level treatment for the mass of people by many assistants. Similar emphasis can be found in the post-revolutionary development of health services in Cuba. In pre-1959 Cuba, the health status of the rural and poor typified that of a people whose government was essentially unconcerned with its welfare. 'Most people in rural Cuba were illiterate or semi-literate, undernourished, and unable to purchase medical care of any reasonable quality' (Guttmacher and Danielson, 1977, 387). There was only one rural hospital (Smith, 1979) and most rural dwellers and blacks everywhere were excluded from the mutualist services which were based on a prepayment system. With unequal access to medical care and with the emphasis of this care on treatment, there were high rates of infections and parasitic diseases, especially amongst the very young. As Guttmacher and Danielson (1977, 387) comment, 'The health profile of the Cuban population reflected its underdevelopment, with infectious diseases overshadowing the chronic diseases whose predominance is commonly considered an indication of socioeconomic and medical development.'

The development of Cuban socialism emphasised moral values, mass mobilisation and central planning (see Ritter, 1974). In the health field, these emphases were reflected in three goals: the centralisation of planning and decentralisation of administration; involvement of other social institutions and of the people in health education and promotion; and the integration of health care facilities. These goals were in turn embodied in a strategy that stressed the development of rural health services.

While there is a spatial hierarchy of services – provincial hospitals and clinics; regional institutions for specialist personnel, support consultation and laboratory facilities; and the community (area) level of the health centre – it is the community level that has received most attention. In the early years after the revolution, this level was developed with the aid of enthusiastic doctors and nurses, lay participation in health promotion and compulsory rural service for graduating doctors. After 1965, the polyclinic – integrated health centre – was emphasised as the first point of access to physician services. As Danielson (1981) points out, these clinics accounted for 32.3 per cent of all medical visits in 1964 but 63.3 per cent in 1969. Linked to the idea of service decentralisation, the polyclinics were primarily outpatient centres, serving populations of around 25,000 people, although in sparsely populated areas this figure might drop to 7,500. The clinics were administratively independent of their associated regional hospitals and were staffed by specialists (internist, paediatrician, obstetrician-gynaecologist, dentist) rather than general practitioners.

Such organisation resulted in user complaints about waiting lists, the problems of hospital visits, the lack of continuity in medical personnel. Indeed, the doctor seemed to be served by the community rather than vice versa, in conformity with a medicine of consumption. A new polyclinic programme was, therefore, instigated in the mid-1970s as part of a large programme called 'medicine in the community'. This programme is centred on doctor-nurse teams, whose work is 'sectorised' as was that of the sanitarians. In other words, the teams are responsible for the health promotion and maintenance of people in a specific area. They are expected to spend some twelve hours a week in home visits and other related community work. The doctors are also expected to participate directly in meetings between patient and specialist. Cuba's rural health system is being based, therefore, on an areal case-work approach to care.

Despite these organisational problems, the health profile of rural Cubans has changed dramatically. Smith (1979) suggests that improvements are more noticeable in the control of conditions, that is improvements in hygiene and sanitation, especially diptheria, malaria and parasitic diseases, than in general or infant mortality. Guttmacher and Danielson (1977) imply that the change is more dramatic, suggesting that the major causes of death have shifted from infectious diseases to chronic illnesses, cancer, stroke and heart disease. They comment further that this change, especially the decline in early childhood mortality, cannot be simply attributed to preventive and sanitary measures. This shift in the cause of death can be seen from Table 6.4, which also shows the great improvement in rural health standards compared with urban areas. Emphasis on rural health is shown in the generally lower rural mortality rates, with rural advantage increasing with age. The reverse of urban-rural differences is quite uncommon for a country with a large rural population and agricultural workforce. Thus, for example, Vallin (1975) points out that the crude mortality rates in Algeria vary from 9 per 1,000 in Algiers, 14.1 in other towns, 16.56 in villages, to 19.8 per 1,000 in the rural areas. Hughes and Hunter (1971) show that for tropical Africa the prevalence of bilharziasis is 15 to 66 per cent of the population, filariasis 20 to 65 per cent, yaws

27 to 50 per cent, leprosy 13 to 53 per cent and malaria 26 to 80 per cent. They suggest that such prevalences reflect a low level of control over nature. But it also reflects the importance attached to the health of the workforce in many societies. Laurell (1981) suggests that the type of agricultural practice (subsistence, wage-labour, etc.), the land-ownership pattern, the role of the market and nature of social relations all affect rates of mortality and morbidity. In Cuba, such phenomena have changed to the benefit of the agricultural workforce and rural population.

Table 6.4: Main Causes of Urban and Rural Mortality in Cuba, 1973 and 1978 (Rates per 100,000 Population)

	1973		1978	
Cause	Urban	Rural	Urban	Rural
Heart disease	151.2	99.1	168.1	124.2
Cancer	93.9	77.3	76.4	67.7
Cerebrovascular disease	60.4	40.6	46.0	40.7
Influenza and pneumonia	56.8	27.8	47.1	36.1
Diseases of arteries, arterioles, capillaries	28.5	27.7	13.9	15.1
Accidents	19.1	20.0	25.6	20.8
Bronchitis, emphysema, asthma	14.7	8.7	7.5	5.9
Suicide	14.6	10.6	7.3	10.7
Diabetes mellitus	13.1	5.6	13.6	6.2
Hypertension	11.1	14.5	8.2	5.6

Source: Laurell (1981), p. 7.

Thus the health profile of rural Cubans is changing. Emphasis has been placed on prevention, hygiene and epidemiology. Little technological innovation has been adopted and increases in expenditure have been moderate. Health care provision is based on the principle of each according to need and on mass participation in health education. The ideas of Guevara (1968) on the importance of education, mobilisation and fulfillment in creating socialist man should not be underestimated, although they have been tempered by practice. But in suggesting that mechanical or technological reality will not dominate in Cuba, Guevara has indirectly shaped the nature of medical education. While it is, therefore, recognised that medical innovation can provide control over nature, it may also result in control over man by an expert group with particular models of man and society. Medical education has therefore been reshaped – aided by the exodus of about one-third of the doctors practising in pre-revolutionary Cuba – so that professional constraints do not militate against other health care aims. Medical education in Cuba strongly emphasises preventive and epidemiological concerns. Medical students are associated with other health workers in work-study programmes and all are made aware of the social meaning of their work. Clinical services which act as teaching settings have developed a strong community bias, which has been aided by the decentralisation of medical training in general away from Havana. Thus medical education as well as health care provision itself reflect in the changing conceptions of health and man in Cuba. These are in turn predicated on the changed social relations brought by the revolution.

Social Justice and Health Care Provision

In this final section, we wish to address the relationships between health care, models of man and society, and social structure somewhat differently. It is a central tenet of this chapter that the nature of health care is based on the nature of the relationships between individuals and groups in society. In other words, it reflects the way different groups see and treat each other. The provision of health care presupposes power relations, particularly important being the differential distribution of power and resources between groups. Indeed the differential command exercised over the use of a society's scarce resources is important in shaping the entire welfare system. We address this issue in the following section using Titmuss's (1976) idea of the social division of welfare. Further, it is possible to argue the shape of the health care system – the outcome – is but a reflection of the struggle for resources, a struggle which can be seen through the moral values of society. In other words, health care provision is essentially a question of social justice, always remembering that notions of justice are implicated in the economic and political realities of their containing societies.

Social Division of Welfare

The allocation of resources to health care and welfare is a power question. The particular dominant interests in a society will not only greatly influence the total expenditure on such services, but where that financial burden is likely to fall. Titmuss (1976) identifies three different types of welfare. All these collective interventions serve both the needs of individuals and the wider interests of society. Thus the social division of welfare 'is not based on any fundamental difference in functions of the three systems . . . or their declared aims. It arises from an organisational division of method which is, in turn, related to the division of labour in complex industrial societies' (Titmuss, 1976, 42). The three 'systems' are social welfare, fiscal welfare and occupational welfare. We have dealt largely with social welfare, that is that based on direct expenditures by the state to ensure the well-being or improve the health status of its citizens for whatever purposes. Social welfare may be provided by selective, universalist or positive discriminatory policies. In other words, it can be found in societies with different power structures, different models of society and different conceptions of health. The balance between social welfare and occupational and fiscal welfare will, however, be governed by societal context and constraints and the role that these constraints allow for publicly-funded welfare.

Fiscal welfare is indirect state aid for the private provision of health care and welfare services, although in most countries it also involves direct assistance to families through family allowances. But in a variety of countries, tax concessions and allowances are available for medical expenses (Australia, Ireland), life insurance contributions (Austria) and superannuation payments (Denmark) (see Kaim-Caudle, 1973; Higgins, 1981). The private market in welfare is best-illustrated by occupational welfare, the sponsorship and health care provision and other benefits by companies, trade unions, or both. Such welfare is especially important in the USA and in Japan. Sinfield (1978) suggests that in the case of the USA it is negative attitudes towards state intervention and positive beliefs about individual freedom, work and private enterprise that ensure the dominance of such a system. The roots of private welfare are thus 'individualism, religious freedom, a relatively free market economy, nonideological unionism and industrial wealth' (Wilensky and Lebeaux, 1965, 161). Individual provision of medical insurance, however, is the dominant route to health care in the USA and occupational wealth is a minor adjunct to this. In Japan the reverse is true.

We pointed earlier to the quasi-feudal relation between

employer and employee in Japan. The implications of this relation can be seen from Dore's (1973) comparative study of the English Electric factory in Bradford and the Hitachi factory in Furusato:

> Hitachi Company's total expenditure on housing, medical services, canteens, transport subsidies, sports and social facilities, and special welfare grants other than pay during sickness, amounted to 8.5 per cent of total labor costs. English Electric figures are not available, but for the median British firm of a group surveyed in 1968 the corresponding figure was 2.5 per cent *including* sick pay.

Occupational welfare schemes though benefit some more than others – those in large firms being especially fortunate. It does mean, as Higgins (1981) points out, that irregularities in the workplace are carried through into inequalities in non-work life.

Private welfare, occupational welfare and fiscal welfare are not solely the preserve of the most free market economies. Such provision is found in most capitalist societies. In some: Denmark, Britain – there has been a reduction in the public expenditure on social welfare – Wilensky (1976) has suggested that this is a reaction against high levels of public spending and taxation and against the speed at which these levels have increased. In the case of Britain, Gough (1979, 140) argues that expenditure is likely to 'switch from direct state provision of services to public subsidisation and purchase of privately-produced services'. This change in the nature of provision, from social to fiscal and private welfare, results in part from the ascendancy of a different model of society, serving re-emerging interests. Thus, in Gough's terms, the reprivatisation of the welfare state is an element in the re-capitalisation of capitalism.

Such a change has important societal implications. Para-doxically, although a change from social to private welfare may increase inequality, it may also enhance social integration. This adjustment in method of provision simply reinforces how the advantaged are able to restrict the visibility of their privileges and subtly pass on the costs to the less powerful. Thus, as Sinfield (1978, 136) argues, 'inequality in the visibility of benefits is an important and integral part of the social division of welfare'. With the ability to pay for private consultations and to queue-jump, the deterrents to medical care identified by Cooper (1975, 16), 'the necessary expenditure of time and energy, inconvenience, travel costs, leisure foregone, the discomforts of the doctor's waiting room', are increasingly passed on to the less powerful. The occupational and fiscal benefits enjoyed by the middle and upper classes further ensures and reinforces social and residential segregation.

The increase in private welfare alters the form in which health care is provided. Tussing (quoted in Higgins, 1981) suggests that it is the form of provision rather than the content of a particular policy that determines whether or not beneficiaries will be stigmatised and that ensures the legitimacy of the models of society underpinning welfare provision. It is those remaining under public welfare schemes who are stigmatised, especially if those provisions are means-tested or selective (see Chapter 2). Tussing points to a dual welfare system – one poorly funded, stigmatised and stigmatising; the other, virtually unknown, non-stigmatised and non-stigmatising. He is essentially describing the division between the public and private systems in the USA, although his comment is of some relevance to all capitalist societies. It certainly reinforces the idea of public welfare as an agent of social and behavioural control.

It is probably fair to say that this view of health care provision is based on the residual model of welfare – a concept of Wilensky and Lebeaux, who identified the residual and institutional models of care. They suggest that:

> The first (residual) holds that social welfare institutions should come into play only when the normal structure of supply, the family and the market break down. The second (institutional) in contrast, sees the welfare services as normal 'first-line' functions of modern industrial society . . . They represent a compromise between the values of economic individualism and free enterprise, on the one hand, and security equality and humanitarianism on the other. (Wilensky and Lebeaux, 1965, 138–9)

Such models imply that there are a variety of health care choices and options open to societies but, given the dominance of such values (and, therefore, interests), certain choices are not given serious consideration. Thus structural and institutional forces are implicated in health care policy choices. We, therefore, concur with Sinfield (1978, 153) who argues that:

> by drawing attention to the importance of power and the

dynamic effect of changes over time in the political economy of the social division of welfare, one is forced to consider again the relation between welfare and the particular form and structure of the society under study.

Health care policy and the spatial allocation of resources are at root questions of power and social justice.

Social Justice and Variations in Health Care Provision

Social justice is quintessentially a practical matter. It is not merely a philosophical issue but 'something contingent upon the social processes operating in society as a whole' (Harvey, 1973, 15). Social justice is a principle or set of principles for resolving conflicting claims over the allocation of scarce resources. We can immediately see that the principle(s) will be related to the nature of power relations and the associated models of man and society. In his analysis of social justice and spatial systems Harvey (1973, 97–8) argues that:

> The principle of social justice therefore applies to the division of benefits and the allocation of burdens arising out of the process of undertaking joint labour. The principle also relates to the social and institutional arrangements associated with the activity of production and distribution. It may thus be extended to consider conflicts over the locus of power and decision-making authority, the distribution of influence, the bestowal of social status, the institutions set up to regulate and control activity, and so on. The essential characteristic in all such cases, however, is that we are seeking a principle which will allow us to evaluate the distributions arrived at as they apply to individuals, groups, organisations, and territories, as well as to evaluate the mechanisms which are used to accomplish this distribution.

Thus social justice involves value judgements about the means and ends of distributional policy, of health care provision. As values vary, principles of social justice will vary. The social Darwinistic values of the market-oriented military dictatorships of Latin America will suggest that it is only 'natural' justice that the richest (the fittest) have the best health care facilities. Thus Escudero (1981) argues that the reemergence of military participation in Argentinian civilian life has led to a lowering of health standards, increases in the rates of mortality and morbidity and the establishment of elitist, commodity-centred health policies. A less extreme but similar view of social justice may be said to underpin USA health care provision.

Other conceptions of social justice contain within them the need to ameliorate the conditions of the disadvantaged. Whether such amelioration is aimed at the reintegration of the less powerful into a market society, the creation of equality of opportunity or the heightening of consciousness for more fundamental change is of course a value question. While we cannot fully address such questions in this book, we wish to take this ameliorative view of justice farther, using the work of Campbell (1978).

Campbell isolated three moral values associated with social justice – freedom, equality and fraternity. He suggests that these are interdependent phenomena and that their separation in the formulation of health care policy leads to injustice. Separation can thus in fact lead to partial or particularistic definition of the values. Freedom, for example, is a central tenet of the American health care system, but it is taken to mean the reduction to a minimum of the constraints on self-interested behaviour. Freedom becomes a goal to be attained through competitiveness and coercion and once achieved to be jealously guarded through the exercise of power. Health care priorities and the spatial and social distribution of facilities and manpower are determined by the purchasing power of the wealthier medical 'consumers', by the influence of the more powerful professional groups and by controls exercised by financial backers. Government actions – medicare, mediaid, food aid – must compete on the terms already established by the entrepreneurs. Such actions and the associated attempts to redefine priorities – as happened in Australia in the mid-1970s – appear as erosions of commercial and professional 'freedom'. Thus, as Campbell argues, it is necessary to temper freedom with equality.

Equality and equity are ambiguous terms as well. Equity suggests the sharing out of benefits and responsibilities proportionately to people's needs and capacities, while treating those with the same needs and capacities equally. Thus equity implies fair discrimination (see Raphael, 1970). The aim of the British NHS is to provide this through equality of opportunity for all irrespective of means, age, sex, class or area and through positive discrimination (see above). Significan inequalities remain between categories of patient, between regions, etc. The determination of the appropriate levels of resources for different areas and services has meant the centralisation and bureaucratisation of decision-making (see above). Such decision-making may only have limited effect, as it is divorced from sick individuals and their changing social contexts and may be greatly influenced by professional and institutional considerations, medical pressure, efficient administration. Attempts to enhance equity may reduce individual and group freedom, especially for less powerful, children from deprived areas, the mentally handicapped and the elderly without adequate family support. Centralised decisions and allocation determine not only equality of access but the type of facility available and when and to whom it is available, as for example in the Soviet Union.

Thus the search for equality in health care needs to be tempered by other considerations. Such tempering may ensure that health care remains a moral end in itself, rather than a political instrument. In fact, centralised systems remove the autonomy of individuals. Campbell uses the Kantian notion of autonomy – the regulation of the self to coexist with the freedom of others – to temper the centralised provision of equity. He suggests that such autonomy is best found in societies which have attempted to restore man as a social being. Thus the social individual must regain consciousness of himself as a self-creating member of the human species, finding fulfillment in productive labour, in the development of higher capacities and in co-operation with his fellows. Autonomy is thus produced by fraternity, which can best be seen in health care policy and the spatial allocation of health care resources in China (see above). The enhancement of fraternity, however, resolves itself into a problem of the allocation and control of political power – the basis of the Trotskyite and Maoist vision of permanent revolution. Fraternity suggests that all individuals are providers and potential recipients on health care, making their assessments and definitions very important (see Chapter 7 for lay concepts). Paradoxically, then, social justice in health care provision may require the centralisation of power to ensure equity and its decentralisation to promote fraternity and co-operation. As Campbell (1978, 71) notes:

> By a curious paradox the provision of nationalised health services in removing people's sense of insecurity about the availability of health service also removes people's sense of

urgency about health care, and so loses the vast potential for self-help which the community possesses.

We must add a rider: such self-help must be based on an understanding of the material world and on the communal provision of health care, unless it is to become 'healthism' (see Chapter 2).

The relative emphases in different societies on fraternity, equality and freedom result in different health care systems. Consideration of social justice enables us to argue more explicitly that health care in one society is not a unique creation, pointing to the importance of comparative analysis (see Chapter 7). Similar values and interests are at work in most societies, albeit pulling in different directions and deriving from different historical contexts. Health and health care are products of social conditions as is social justice. And like social justice, they can modify those social conditions. We are dealing, therefore, with variations on, and variants of, similar themes – both social justice and health care provision exemplify the dialectic between ideas and the social order (see Eyles, 1981b). To particularise, there is a dialectic between conceptions of health and policies of health care and the social, economic and political structure.

This dialectic has a spatial outcome. We can utilise the work of Smith (1982) to demonstrate this outcome. In fact, it is possible to suggest that the patterns of spatial organisation and service provision are reflections of social structure and conceptions of health. Smith uses the examples of the USA, Britain, USSR and China as they might be seen in a segment of geographical space from the centre of a large city out into the countryside (Figure 6.3). An important dimension of differentiation is again the public-private distinction. The USA exemplifies an almost completely private (fee-for-service) system with low public provision. There is a complete spatial hierarchy for those with the ability to pay with the care of the poor often being limited to one public hospital in each city. The British case illustrates a national health service with a (growing) private sector. There is a comprehensive universal spatial delivery of care, although this is limited by past form and structure and public expenditure constraints. The Soviet Union has a full public hierarchy. Inequalities arise from urban-rural differences and from elite health care privileges. China exemplifies a spatial diffuse, prevention-orientated system, based on low-

**Figure 6.3:   Different National Patterns of the Spatial
Organisation of Health Care Facilities (Generalised)**

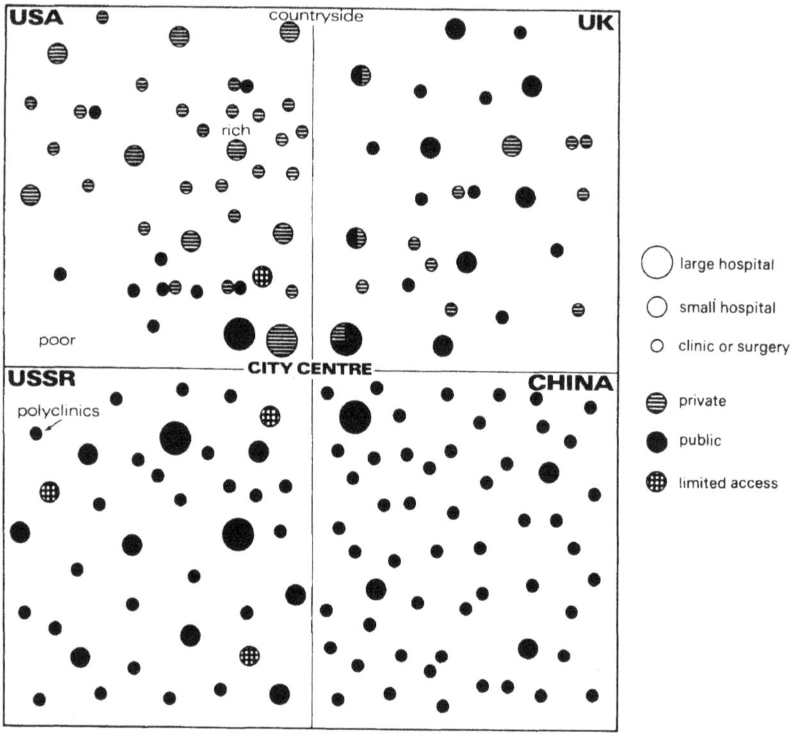

Source: Smith (1982), p. 9.

technology mass participation and limited centralised facilities.

The parallels between spatial form and social process and conception seem fairly obvious but nevertheless worth demonstrating. These parallels add further weight to our contention that health and health care are societal products. Indeed, it is possible to suggest a sequencing of phenomena, from social structure and social relations to views of man and of society to views of health to the provision of health care to the spatial delivery of that care. With such sequencing we are not trying to suggest that there is a simple progression from one end to the other. There is certainly feedback. It is most likely that we are dealing with a multiple linkage system in which all elements affect and constrain all others. A change in one element, whether from an internal or external source, is likely to produce changes in all other elements.

Thus the bureaucratisation and professionalisation of health care provision in the advanced industrial nations has not only influenced our conceptions of health but has had an impact on the spatial delivery of care. Health has become a curative, technologically-orientated matter and is seen as being best catered for in centralised facilities; at the primary level the health centre instead of the one doctor general practice; at the hospital level, the district rather than the 'cottage' facility. But having suggested that the sequence is an interactive one, we would argue that certain parts are more important than others, the most vital being social structure and relations and the conceptions of man, society and justice emanating from these.

In this final chapter, we wish to draw together some of our contentions and conclusions and suggest, tentatively, how the understanding of medicine and health may be advanced. Throughout the book, when considering disease, medicine, health care and health itself, we pointed to the role of the social fabric. The natures and effects of all these phenomena are, therefore, shaped and influenced by their containing societies. This is not to suggest that their impacts or effects on such factors are precisely the same for all individuals and groups within those societies. Other factors – environment, work, class, culture, personality – intervene to predispose individuals and groups in specific ways. Health inequalities and differential ecological association between disease and group are but two examples of the influence of such factors. Nor should we suggest that health, medicine and health care are simply reflexes of the social order. We cannot determine the conception of health or the nature of the health care system simply by looking at the nature of societal development. Predisposing factors do not operate uniformly, nor always in the same direction. Once formulated, conceptions of health and health care delivery systems have their own impacts on the social order. There is thus no unique logic of industrial development and the intended consequences of health care provision may produce some unintended effects. Individuals and groups, for example, do not simply accept their fate; or, more accurately, they do not acquiesce for ever. They learn from their own experiences – from their own health status and own accessibility to health care resources. Their experience, therefore, contains within itself an immanent critique, not only of everyday life but also of the institutions and structures that shape that life.

While it would be easy to look retrospectively for examples of the foregoing arguments in the earlier chapters, we wish rather to look forward to point to possible ways of discovering more about such relationships. Indeed, we see as central the necessity to understand better the relationships between medicine and health on the one hand and societies, individuals and environments on the other. In suggesting ways of looking at these phenomena and their interrelationships, we recognise that we are saying nothing really new. We thus see a need for both macro- and micro-analyses – analyses that focus on society and environment and on individuals – as well as studies that see individuals and society as a dialectical whole. In fact, we argue that it is only through understanding such a whole – focussing on both societal structures and institutions and individual and group experience – that we can gain a deeper knowledge of the nature of social change.

We intend to focus on five topics: comparative analysis; institutional analysis; the relationships between models of society and health care; lay concepts of health and illness; and the nature of change.

Comparative Analysis

In the context of sociology, Durkheim (1938, 139) argued that 'Comparative sociology is not a particular branch of sociology; it is sociology itself, insofar as it ceases to be purely descriptive and aspires to account for facts.' In other words, it is the method of social science as it enables generalisations to be made. These generalisations result from the method entailing the study, side by side, of different groups, collectivities, institutions, communities and environments which appear to present some important similar features. Thus, through comparative analysis it may be possible to move away from ethnocentric, particularistic explanations. While particularistic explanations are important for understanding developments in any one society or environment (see below), such possibilities suggest why comparative (or cross-cultural) studies are so important in social anthropology. Higgins (1981) argues that such a method is vital for a fuller appreciation of social policy. She suggests that it encourages a distinction between the general and specific, approvingly quoting Heclo (1972, 95) who comments that:

> to speak of comparative analysis suggests not only that one will be looking at variables which actually vary, but also that one will be doing so in contexts which themselves vary . . . it is only through such comparative analysis that one can appreciate what are the truly unique and what are the more generic phenomena.

Further comparative analysis of health and health care allows for the recognition of alternative courses of action (see Chapter 6). It can also identify and evaluate fashions in social policy and social determinants of policy. In other words, are particular health definitions and delivery systems culturally or systemically conditioned? Are the natures and relative locations of health facilities products of the mode of environmental adaptation (culturally conditioned) or of the nature of economic and political relations (systemically conditioned)?

Such analysis provides, therefore, a broader perspective, potentially countering the problems of single society studies. Indeed, Carrier and Kendall (1977) point to four such problems. First, there is the 'fetish of the single cause', approaching the explanation of health care development from a single aim or single cause perspective. The geographical search for pathogenic environments and Piven and Cloward's (1970) view that American poor and sick relief was aimed overwhelmingly at social control of marginal labour would exemplify such explanation. Illness and health policy responses are usually more complicated than that. Secondly, history may be seen as 'hagiography and biography', that is an overemphasis on individuals and the institutions they created, modified or resisted. Thirdly, there is often a search for 'turning points' in which legislative landmarks or key dates are identified. In the medical geography literature, there is a danger that John Snow and his role in identifying the relationship between cholera and infected water in 1854 may be so viewed; as may 1948 as the beginnings of the NHS. Fourthly, there may be a tendency to 'grand' and '*a priori*' theorising on the basis of limited evidence. The idea that welfare states are an inevitable outcome of industrialism would be an example.

It may be apposite to provide a brief example of comparative analysis. Titmuss' (1973) study of the gift relationship, of blood donors, has both health and geographical significance. It examines comparatively the issues of freedom of choice, uncertainty and unpredictability, quality, safety and efficiency and effectiveness in relation to the supply and distribution of blood. It thus looks at the variations between nations for this element of health policy and explores the dependence of the sick on fellow citizens. Titmuss found great variation in practice with Britain and Eire offering no payment to donors while Sweden and Egypt paid all donors. Japan, West Germany and the USA paid most donors. These data enable Titmuss to examine the relative efficacy of private and public provision of health care. He found that the private market was wasteful and led to shortages. In the USA, for example, the reliance on blood supplied for cash or in return for past or potential supply means great difficulty in securing adjustments of supplies to short-term changes in demand in any locality. Thus, hoarding (despite the 21 day life of stored blood) and wastage (because of this shelf life) by hospitals are common. Further, competition has not produced better standards of provision and care. Paid donors are more likely to supply contaminated blood, as many depend on such payments for their livelihood. Such blood is also often contaminated by the donors being drug addicts or alcoholics or suffering from hepatitis, meaning that patients receiving blood from paid donors face greater risks from serum hepatitis than in, say, the British system where donation is motivated by altruism.

Titmuss' study also demonstrates the social determinants of policy. When unpaid and voluntary, donors are motivated by altruism, reciprocity and duty. There appears to be a consensual view of man and society underpinning such giving. We must remember that the opportunity cost of blood donation is low (Collard, 1978) and that the need for the service and its technical definition seem unproblematic (Room, 1979). We must be wary of automatically applying the models to other areas of health policy. In the USA and Japan, there is a contractual relationship between donor and hospital and patient because of the financial rewards, while in the USSR, Latin America and parts of Africa donors are motivated by the prospect of benefits in kind and special privileges. Overall, the differences between societies appear to suggest that social context greatly influences individual behaviour and health outcomes – a central contention of our book.

As we suggested, our topics are not new. The work of medical geographers on, for example, environment and oesophageal cancer (McGlashan, 1982) is a comparative study. Nor is comparative analysis without problems. Higgins (1981) suggests that comparative analyses have often been untheoretical and have failed to take sufficient account of different definitions of terms like 'policy' and 'welfare'. Carrier and Kendall (1977) point to the difficulties of cross-national data availability and interpretation; of selecting accurate and adequate indices of contexts and change; and of separating correlation and causation. They add that 'the purposive social action of the members of society – their *understanding* of industrialisation, *their knowledge* of social problems and *their ideas* about welfare development . . . – are excluded from analysis' (1977, 287) and hence from an understanding of causation. As these problems are particularly acute in comparative research, they suggest that while accurate accounts are almost impossible to obtain, a set of plausible accounts of welfare developments may emerge. These 'accounts' can be seen as cognitive constructs related to the behavioural-phenomenological perspectives outlined in Chapter 1. Although we were primarily concerned there with the social construction of diagnoses and illnesses, we can readily see that policy is also 'socially constructed'. It too is the interpretation of events and process by individuals and groups, particularly decision-makers. An investigation of resource allocation policy changes would be instructive from such a viewpoint. As a brief suggestion, it is possible to argue that Room's (1979) discussion of the 1974 reorganisation of the NHS is in this tradition. He regards the policies of the Convervative and Labour parties as being based on different 'readings of the past'. In other words, they were based on different socially constructed interpretations of past events. With such considerations we are no longer looking at comparative analyses but those of particular societies. It is to such analyses that we now explicitly turn.

Institutional Analysis

We can learn a great deal from the analysis of particular societies or environments, especially if we attempt to get beneath surface phenomena and discover the underlying forces and processes at work. Such an aim would suggest a materialist perspective, although we would emphasise that a fuller understanding of phenomena and processes comes from a broader consideration of that provided by orthodox marxism. We have shown already how Navarro's (1978) attempt to explain the development of the NHS in terms of capitalist relations and economic struggle founders (Chapter 2). A broader perspective would take account of power blocs, class alliances (say between radical doctors and working-class groups), the role of the state and the shifting relationship between hegemony and experience. We wish to show the relevance of such a broad perspective by examining the changing nature of spatial resource allocation in London's health services (see Eyles *et al.*, 1982).

At first sight, the provision of health services in London appears to be simply a question of allocating resources within and between territorial units defined for administrative purposes. It appears as a technical problem capable of rational solution (see Chapter 4). But by employing apparently sophisticated scientific analysis, the administrative apparatus of the health service gives the appearance of having devised a rational allocation procedure for NHS resources. In the case of London's health services, the procedure has been largely restricted to the acute hospital sector, reflecting the control over resources exercised by this sector and its staff, the most technically and scientifically-oriented part of medicine. Further, it is possible to argue that in the NHS, health care planning has become dominated by the needs of the service rather than the needs of the population. Bureaucratic administration can take on a life of its own, seeing its own aims and interests as dominant. These aims, specifically the rationalisation and routinisation of events, are best served by adjusting existing levels of provision and distribution of supply with little regard to the health needs of local populations and to the suitability of existing resource mixes to meet them. Indeed, it may be impossible for the health services to have more than 'little regard', because of the structural constraints under which all state agencies work.

Thus, rationality and technicalism have societal implications. Rational, scientific appraisal is one of the bases on which the state and its agencies can claim to act as independent arbiters between competing territories and groups. The assumption of independence means that the political authority of the state appears not directly reducible to power derived from the social structure. The state is thus seen as existing outside or above the social order, intervening in administrative rather than political ways. Further, in the context of the NHS and London's health facilities, expenditure constraints have been applied. Thus, while it is possible to claim that equity criteria have been adopted, the rational procedure does in fact limit the size in the rate of public expenditure because the 'under-funded' gain at the expense of the 'over-funded'. The released funds may be available, through lower tax burdens, for greater private investment and the better realisation of capital. Institutional activity may thus be contradictory. On the one hand, the state through its agencies and bureaucracies intervenes to improve the infrastructure (health, education, transport, etc.), to provide the conditions for the efficient and profitable running of industry and the service sector which under capitalism are, by and large, privately owned. On the other hand, such intervention, by expanding (spending more on) the public sector, appears to threaten the existence of the private sector and hence of capitalism itself. It becomes necessary, therefore, to renunciate and cut back public investments to release funds for industrial (private, capitalistic) development.

A consideration of spatial resource allocation procedures has led, therefore, to a questioning of the rational and technical criteria on which such allocations are based. This questioning is not only a technical matter but an institutional and systemic one as well. It suggests that the relationships between state and economy and between the various state agencies and practices are not irrelevant to an examination of health care delivery. Nor too are social relations if Walters (1980, 160) is correct in suggesting that:

> the NHS represents an attempt on the part of the state to rationalise the organisation and delivery of health care. In so doing the state defined health largely in terms of access to care and by claiming to provide universal access to care, it has contributed to the belief that class inequalities in health and access to care have virtually disappeared.

To explicate these relationships and enhance our understanding of health and health care, different notions of the state may be required (see Chapter 6). In any event, such institutional analysis does demonstrate the close association between health and society. It also shows how a consideration of geographical variations in the allocation of health care resources may lead to an examination of state practice and of the interactions between state and economy. We would argue that such analysis shows not the irrelevance but the strength of geographical approaches to the study of health care. Indeed, health care policy must have a geographical component in that resources are allocated on a territorial basis. Such allocation is likely to remain as the various aspects of medical care are effectively directed at populations in areas, as the Cuban experience shows (see Chapter 6). A greater understanding of state and bureaucratic practices is likely to feed back into spatial allocation policies. The mutually interactive nature of health–health policy–society relationships is thus demonstrated. Indeed, it is important to note that while systemic constraints limit health care developments, they may also enable others to take place. And the constraints themselves may be shaped by health care practice and experience (see below).

Models of Society and Health Care

We have suggested throughout this book that the nature of health care provision is both shaped and affected by social structure and the models of society emanating from that structure. We do not intend to repeat the points already made. All we wish to do in this section is make explicit three major models of society. Thus we suggest that there is not only one way of viewing society at any one time. We do accept the contention that there is a dominant model of society or meaning-system (Parkin, 1971) which significantly shapes the lives and experiences of people in any one society (see below). What we identify here are perspectives of the nature of society which may at any time and in any place be put into practice, that is become part of social life itself – ordering and shaping everyday life – but which act also as theoretical constructs against which the operation of society and the health care system may be judged. For the development of such constructs we turn to Room (1979).

Room identifies neo-marxist, liberal and social democratic 'interpretations of social policy'. Such interpretations may in themselves be regarded as phenomenological as they are reading events and processes from specific viewpoints. As such, they may become value commitments that influence the conception of problems and the interpretations of findings. The neo-marxist interpretation, for example, sees conflicting class interests, capital accumulation and the revolutionary role of the proletariat as significant phenomena. Such an interpretation has influenced some of our contentions in this book. We regard conflicting interests as important, contend that health expenditures are related to the power of capital and the role of the state and suggest that hegemony may limit the actions and aims of the relatively powerless. We have used such interpretation, in the main, as critique – as a way of challenging assumptions and views about the world. We accept that this interpretation is predicated on the idea of equity but also recognise that problems – the overcentralisation of planning, elite manipulation of resource allocations – have arisen when equity criteria appear to dominate policy (see Chapter 6).

Secondly, there is the liberal interpretation. Room (1979) divides the liberals into two categories. The market liberals see the competitive market as crucial in granting freedom in the disposition of individual skills and property. Social policy must support and reinforce the operation of the market. It must not be a burden on economic activities. In fact, welfare bureaucracies are cost-inefficient and their activities should be limited to that necessary to provide a safety-net for social integrative purposes and to ensure that 'human investment' is organised in the interests of economic productivity. It is possible to interpret both the American and Soviet public health care systems in this way. Political liberals are concerned to minimise social conflict through the use of widely agreed procedures. Their primary aim is to help promote identity of interest among social groups. The state is seen as a benign arbiter, helping to reintegrate the labour force into society by ensuring its standards of living and health. The state simply has to do this – the logic of industrialism argument – to sustain the workforce necessary for an advanced society.

According to Room, the social democratic interpretation is based on the societal recognition of a 'needs nexus'. Social policies are therefore directed at the needs of individuals and are not necessarily based on their economic power. Further, social policy is formulated on the basis of choice: support must be obtained from the population at large which is able to put its experience to good use, influencing policy so as to improve its life conditions. Such influence is based not on self-interest but the recognition of being part of a moral community as a citizen. As Room recognises, such an interpretation underemphasises concentrations of power and the acquisitive individualism that dominates many capitalist societies.

The use of one of these interpretations depends of course on the value position of a researcher and his commitment to a particular form of society. Thus a major purpose of this section is to reiterate that in studies and analyses of welfare and health care, value judgements cannot be avoided. All decisions about what constitutes health and the nature of health care delivery have value connotations. Ethical judgements are involved. The use of particular interpretations or models of society is simply another way of implying the same. These interpretations suggest a commitment to a particular form of social life – the neo-marxian and social democrat to a more equitable society although the former vehemently rejects the latter's acceptance of the existing framework of economic and political system; the political liberals to a pluralistic society in which interest groups coalesce to bargain for resources; and the market liberals to a society of self-interested individuals, free to pursue those interests as far as their skills and property might take them. While these interpretations are not always clear-cut in practice, we can identify policies that have been formulated under their guidance. What we need to make more explicit is their use as theoretical constructs in the assessment of health care policy. We do not regard such assessment as introspective. Understanding increases when value-premises are known. Thus we support Carrier and Kendall (1977, 288) when they argue that 'it is our contention not only that any one set of "known facts" can generate a range of explanations but that the "known facts" . . . are themselves a variable and therefore there can be a range of plausible "known facts"'. That plausibility will depend of course on the interpretation of society adopted.

Lay Concepts of Health and Illness

An often neglected dimension of health and health care from the range of 'known facts' is that of lay concepts. Such concepts may be regarded as aspects of illness behaviour (see Chapter 5) – the micro-analytical dimension of health studies which concentrates on the individual and his immediate setting. In other words, structural and institutional phenomena will be considered in so far as they affect individual and small group behaviour.

Cultural and historical variations in such behaviour are important because they present an additional dimension to the studies of health care provision discussed in Chapter 6. Thus in the past in both Britain and the USA it may have been certain forms of psychoses that were labelled as witchcraft. Robinson (1971) suggests that tiredness is viewed differently by the middle and working classes in Britain. The former group see it as an illness while for the latter it is accepted as a normal part of everyday life. In Dade County, Florida, Johnson *et al.* (1962) found that polio vaccination was readily accepted by members of higher social groups and those who belonged to social organisations. These individuals had more friends and acquaintances and interacted with these associates more frequently than other people in the locality. The hard to reach and those unwilling to recognise the risk of illness were often educationally and occupationally disadvantaged. They reported fewer group memberships and less frequent contact with friends.

It is quite easy to discern differences in illness behaviour between different cultures. Zborowski (1953), for example, noted that Jewish and Italian patients in New York City reacted to pain in an emotional way, while native Americans were more stoical and the Irish often denied the existence of pain. He suggests that the Jewish and Italian responses are based on their experiences during childhood when their mothers were overprotective. Zola (1966) points out that coughing, sweating and diarrhoea are regarded as normal, inevitable events among Mexican-Americans. Littlewood and Lipsedge (1982) point to the different ways in which depression is viewed in different societies.

Ill-health is, therefore, not simply a matter of the physical environment and micro-organisms. It also concerns those social relations and psychological characteristics that permeate an individual's life history. Health and illness behaviour are to be viewed in the context of the whole of social life (see Hinkle and Wolfe, 1957). As Mechanic (1978, 263) suggests:

> cultural and social conditioning thus plays a major though not an exclusive role in patterns of illness behaviour. Ethnic membership, family composition, peer pressures, and age-sex role learning to some extent influence attitudes towards risks, the significance of common threats, and receptivity to medical services.

These factors, thus, influence response to illness – and we now turn to a consideration of individual interpretation and, therefore, lay concepts of health and illness.

With this second theme, we again emphasise the importance of the behavioural-phenomenological perspective. Parsons' (1951) identification of the sick role as the response to illness has been critically received. Health is related to the performance of social roles. Illness leads to the interruption of normal functioning and the adoption of the sick role, which is the way society controls those released from their usual social responsibilities. The patient is expected to get well and if necessary seek technically competent help to do so. While Parsons has been criticised for providing a view of illness as treatable disease, for overemphasising the benevolence of the technical interventions of medicine, (see Ehrenreich, 1978) and for a Western, work ethic-oriented view of illness (see Dingwall, 1976), he does see ill-health as a role as well as a condition. In other words, it is a social relation. Crude application of the idea of sick role will omit the variations in responses to illness, but the concept itself does point again to the inextricable linkage between health and society.

In many ways, individual responses are not apparently as societally determined as the sick role implies. As Kohn and White (1976, 2) argue, 'a person's responses to ill-health are regulated by his perceptions'. These perceptions are in themselves shaped by experience and socialisation, the nature of which determine, in good measure, lay conceptions. Family and other kin are vitally important agencies in shaping these conceptions. These agencies form the central element in informal social networks which help define illness and which may often provide support in times of ill-health. Informal support networks help moderate stress (see Cobb, 1976; Pilisuk and Froland, 1978). Problems and stresses are seen as less serious threats to health if strong social support is present. Such networks can also act as buffers between the individual and the deleterious influences. They, therefore, help with coping, by providing assistance in times of crisis, as in bereavement (see Kaplan *et al.*, 1977), or by legitimising role exemption (see Shuval *et al.*, 1970). Frequent interaction between members of a social network means that they can keep an eye on each other. It is no surprise, therefore, that the most vulnerable individuals are the marginal, the isolated and those with changing or ambiguous role structures and with inadequate social support (Totman, 1979).

Such informal social networks are also the bases of lay referral systems. In other words, they influence the nature and timing of help sought for different illnesses. Some networks – those of the middle class – share many of the values of the medical profession and often recommend early formal consultation. In contrast, Suchman (1965) found that the more ethnocentric and socially cohesive ethnic groups were skeptical of professional care and were likely to seek self-administered or folk remedies. In general terms, Freidson (1960, 377) has argued, '. . . the whole process of seeking help involves a network of potential consultants, from the intimate and informal confines of the nuclear family through successively more select, distant, and authoritative laymen, until the "professional" is reached'. Consultations do not necessarily follow this ordered sequence. Indeed, the vast majority of help-seeking cases remain at the level of self-diagnosis or referral to the social network (Eisenberg and Kleinman, 1981).

It is, however, still not that clear how informal networks operate, let alone how they operate in the context of specific illnesses. While there have been many examples of social network analysis to discover patterns of interaction and influence, little work has been carried out on the relationships between networks and definitions of, and responses to, illness. Most of that which has been carried out has worked backwards from hospital or surgery rather than starting with networks themselves. A major reason for such starting-points is the difficulty of identifying meaningful networks for the study of illness responses. It is likely that such networks are localised in terms of either geographical space or contact points. It is our view that the identification and isolation of these networks is an eminently social geographical project, although one that will link social geography even more closely with broad social science concerns. Such projects, emphasising as they would, the local production of meanings and interpretations, the relative influences of national and local meaning-systems, the mediations of culture and work on the definitions and practices of health and illness, focus attention on the dialectic between ideas, meaning and interpretations and actions, policies and practices – a central concern of the philosophical debate and of considerations on the nature of change.

The Nature of Change

Discussing the nature of social and the concomitant geographical change is a topic worthy of a monograph in its own right. It is necessary, however, to state briefly how change in medical and health practices may be considered. Many of the analyses of change in social geography take the form of static pictures taken at two or more points in time. Such perspectives are understandable given the nature of much social geographical data. Comparisons are made between two sets of census data or two surveys. While these comparisons tell us how much change may have occurred between one time and another, they seldom inform on how and why such change occurs. Such questions demand a consideration of the nature of society and of social relations. In turn, this consideration requires a model of man and of society at least as a theoretical construct. Thus the how and why questions are value-loaded and demand a commitment to a specific form of society against which the amount of change (the how much question) can be assessed. We do not wish to imply that we have discovered some major omission in the bulk of social geographical work. Such issues were part of the social relevance discussions in the mid-1970s and have, in general terms, been addressed in human geography specifically by Harvey (1973) and Smith (1977). We would further contend, however, that models of society are implicated in much social geographical work. Indeed, one such model is particularly so implicated. It is the acceptance of incremental improvement in the human condition as a moral right of being human. It has much in common, therefore, with the social democratic interpretation outlined above. It is, of course, possible to argue that such a position is both politically and morally responsible, although it tends to underemphasise resource conflicts. Indeed, it further underemphasises the ecological limitations of resources which are likely to exacerbate such conflicts and greatly shape the nature and direction of change (Eyles, 1981b).

To understand better the nature of change it may be useful to see events and phenomena in the process of becoming. In other words, it is possible to see the world in terms of relationships between phenomena rather than in terms of the phenomena alone. Such a view means more than saying that phenomenon A and phenomenon B coexist, it is to see them implicated in one another, to see them in dialectical relation. Such a relation is not a metaphysical category but a concrete, historical form which presents social reality as a structured totality (see Sayer, 1979; Eyles, 1981a). We have both explicitly and implicitly been concerned with two such relations. To clarify, our view of social relations as a structured totality has developed primarily from a consideration of two major perspectives with respect to health and medicine. First, there is the relation between conceptions or ideas and policies or actions. If we wish to understand why particular policies have been adopted, it is not enough to compare policy options. We also have to discover how such policy options were produced. What sets of ideas were available to inform decision-makers? Further, what interests have been supported or disadvantaged by such ideas? Thus the first relation is predicated on the second – that between individual and group experience and institutional and societal structure. While such predication is vital if the social world is to be seen as a structured totality, the second perspective is specifically concerned with how health institutions and structures shape and are shaped by the experiences of specific groups.

These interests – derivatives of experience – influence and, again, are influenced by sets of ideas. The nature of these relationships (and, therefore, the direction and pace of change) seem, in large measure, to be based on the form of social relations – that is those between classes, power blocs, group alliances – and the maintenance of such forms through coercive or hegemonic constraints. We would argue that there is a set of ruling ideas – hegemony, dominant meaning-system – which operates to limit ameliorative and radical change. This meaning-system does not completely dominate disadvantaged and subordinate groups. Their experience, shaped by work, community and family, suggests that there are other ways of viewing the world. It is thus from this experience and its related ideas – which are, in essence, articulated demands – that certain forms of change may develop. Thus sick clubs and friendly societies were derived from the work experience of specific groups in the nineteenth century. Further, the interests of the dominant groupings may require the instigation of specific policies. This relationship is again mediated by the articulation of those interests as a set of ideas and options (see Eyles, 1981b). We refer again to Kincaid's analysis of the Beveridge reforms (see Chapter 2). In Bismarck's Germany the humanitarian notions of welfare and health care had specific interests to protect. As Briggs (1961, 249) argues:

> Many of Bismarck's critics accused him, not without justi-
> fication, of seeking through his legislation to make German
> workers 'depend' upon the state . . . 'Welfare' soothed the
> spirit, or perhaps tamed it . . . Bismarck was anxious to make
> German social democracy less attractive to working-men. He
> feared 'class war' and wanted to postpone it as long as possible.

Finally, the effects on health care structure of the relationships between working-class experience and ideas and dominant interests and policy-making is demonstrated by Navarro's analysis of British health insurance and Smith's study of the black lung movement in West Virginia (see Chapter 2).

If we were asked to generalise on the effects of all the above relations on change in medicine and health, we would have to answer that not only are such relations historically specific but also that we could not. We suggest that this is a set of topics requiring more social geographical research. The starting points of individual and group experience, policy options, specific spatial resource allocation practices or whatever are equally valid – but such phenomena must be seen as implicated in a structured totality of social relations.

Conclusion

We hope to have demonstrated in this final chapter some of the potentialities for social geographical research in health and health care. We make no apology for being selective. Selection is of course biased towards our own particular interests. We also believe that an exhaustive list of geographical projects to be counter productive. Indeed, as Shannon (1980) avers, a disciplinary perspective may be a major hindrance to health research. We have concentrated on the social, seeing explorations of the health–health care–society relationship as central to understanding the social realities of the advanced industrial nations and the health promotion strategies of many Third World countries. We have mainly approached 'the social' through the behavioural–phenomenological and materialist perspectives, seeing 'the geographical' as a pertinent contribution to these routes to understanding and explanation. We do not regard the more 'spatial' aspects of the geographical perspective as irrelevant or unimportant. Analyses of the relationships between disease, medicine and environment and those between resource allocation and territorial structure contribute significantly to our comprehension of health. Medical ecological and epidemiological studies, perhaps employing the biomedical perspective, are particularly important in the developing world, while spatial health organisation work, utilising possibly a behaviourally modified economic approach, is useful for furthering our knowledge of the problems of health care delivery.

Our own approach has ranged widely away from the traditional concerns of health *per se*. We have felt it necessary to examine, for example, aspects of social policy, welfare, poverty, science and technology. All these phenomena are implicated in the relationship between health and society. Indeed, this relationship must come to the fore if society is seen as a structured totality with health as a part of this whole. But the relationship is not one-way – it is dialectical. The same can be said for all the relationships we have examined: disease–environment, medicine–science, environment–medicine, health–culture, health care–society, for example. We hope that we do not conclude on a note of confusion when we say all implicates all. We have tried to suggest key points of entry to this totality – health experience–health structure; conceptions of health–nature of health care. But it is only through constant recognition of this holism that we keep in mind the ever-present relations between, on the one hand, medicine and health and, on the other, other forms of welfare, work, residence, culture and social formation, or society in general.

BIBLIOGRAPHY

Abel-Smith, B. (1964) *The Hospitals 1800–1948*, Heinemann

Abler R. Adams J.S. and Gould P. (1971) *Spatial Organisation: The Geographers View of the World*, Prentice-Hall

Adorno, T.W. (1974) *Minima Moralia*, Heinemann

Advertiser, The Adelaide Australia

Alland, A. (1970) *Adaptation in Cultural Evolution: An Approach to Medical Anthropology*, Columbia UP

Aron, R. (1970) *Main Currents in Sociological Thought*, vol. 2 Penguin

Aronson, N. (1978) 'Review of B.A. Weisbrod *et al.*, Disease and Economic Development', *Social Science and Medicine, 12C*, 66–8

Ashworth, W. (1954) *The Genesis of Modern British Town Planning*, RKP

Australia (1973) *A Community Health Programme for Australia*, Report from the National Hospital and Health Services Commission: Interim Committee, Chairman Sidney Sax, Australian Govt Publishing Service

Avery Jones, F. (1976) 'The London Hospitals Scene', *British Medical Journal, 2*, 1046–9

Avineri, S. (1968) *The Social and Political Thought of Karl Marx*, Cambridge UP

Bagley, C. and Jacobson, S. (1976) 'Ecological Variation of Three Types of Suicide', *Psychological Medicine, 6*, 423–7

Bagley, C. Jacobson, S. and Palmer C. (1973) 'Social Structure and the Ecological Distribution of Mental Illness, Suicide and Delinquency', *Psychological Medicine, 3*, 429–38

Baker, J. (1979) 'Social Conscience and Social Policy', *Journal of Social Policy, 8*, 177–206

Banerji, D. (1975) 'Social and Cultural Foundations of the Health Services System of India', *Inquiry, 22*

Barker, D.J.P. (1982) *Practical Epidemiology*, Churchill-Livingstone

Bastide, R. (1972) *The Sociology of Mental Disorders*, Routledge and Kegan Paul

Bates, E.M. (1980) 'A Consumer's View of the Australian Experience in Health Insurance', *The Lancet*, 5 July, 26–8

Bell, C. and Newby, H. (1976) 'Community, Communion, Class and Community Action' in D.T. Herbert and R.J. Johnston (eds), *Social Areas in Cities*, vol. 2, Wiley

Bell, D. (1960) *The End of Ideology*, Free Press

Berger, P.L. and Luckmann, T. (1967) *The Social Construction of Reality*, Doubleday Anchor

Bevan, A. (1952) *In Place of Fear*, Heinemann

Bevan, G.H. Copeman, J. Perrin, J. and Rosser, R. (1980) *Health Care Priorities and Management*, Croom Helm

Biderman, A.D. Louria, M. and Bacchus, J. (1963) *Historical Incidents of Extreme Overcrowding*, Bureau of Social Science Research

Birch, H.G. and Gussow, J.D. (1970) *Disadvantaged Children: Health, Nutrition and School Failure*, Harcourt, Brace and World

Blaxter, M. (1976) 'Social Class and Health Inequalities' in C.O. Carter and J. Peel (eds), *Equalities and Inequalities in Health*, Academic Press

Boal, F.W. Doherty, P. and Pringle, P.G. (1978) 'Social Problems in the Belfast Urban Area', *Queen Mary College, Department of Geography, Occasional Paper, 12*

Boggs, C. (1976) *Gramsci's Marxism*, Pluto Press
Booth, A. (1976) *Urban Crowding and Its Consequences*, Praeger
Bradley, J.E. Kirby, A.M. and Taylor, P.J. (1978) 'Distance Decay and Dental
    Decay: A Study of Dental Health Among Primary School Children in
    Newcastle-upon-Tyne', *Regional Studies*, *12*, 529–40
Bradshaw, J. (1972) 'The Concept of Social Need', *New Society* 30 March, 640–3
Brenner, M.H. (1979) 'Mortality and the National Economy: A Review and the
    Experience of England and Wales 1936–76', *The Lancet*, 15 September, 568–73
—— (1981) 'Unemployment and Health', *The Lancet*, 17 October, 874–5
Briggs, A. (1961) 'The Welfare State in Historical Perspective', Archives
    Européenes de Sociologie *2*, 221–58
British Medical Association (1978) 'Priorities in the NHS: The Way Forward.
    Statement by the BMA Council', *BMJ*, 1A, 254
British Medical Journal (1982) 'Medical Advisory Machinery in the Re-organized
    NHS', *BMJ*, *284*,64–7
Brotherston, J. (1976) 'Inequality: Is it Inevitable?', in C.O. Carter and J. Peel
    (eds), *Equalities and Inequalities in Health*, Academic Press
Brown, E.R. (1976) 'Public Health in Imperialism: Early Rockefeller Programs at
    Home and Abroad', *American Journal of Public Health*, *66*, 897–903
Brown, G.W. Bhrolchain, M.N. and Harris, T. (1975) 'Social Class and Psychiatric
    Disturbance Among Women in an Urban Population', *Sociology*, *9*, 225–54
Brown, G.W. and Harris, T. (1978) *The Social Origins of Depression*, Tavistock
Brown, P. (1973) *Radical Psychology*, Harper and Row
Burgess, E.W. (1939) 'Introduction: R.E.L. Faris and H.W. Dunham', *Mental
    Disorders in Urban Areas*, University of Chicago Press
Burnley, I.H. (1977) 'Mortality Variations in an Australian Metropolis: The Case
    of Sydney', in N. McGlasham (ed), 'Studies in Australian Mortality', *University
    of Tasmania, Environmental Studies, Occasional Paper*, *4*.
Butler, J.R. Bevan, J.M. and Taylor, R.C. (1973) *Family Doctors and Public
    Policy*, Routledge and Kegan Paul
Butler, J.R. and Morgan, M. (1977) 'Marital Status and Hospital Use', *British
    Journal of Preventive and Social Medicine*, *31*, 192–8
Buttimer, A. (1969) 'Social Space in Interdisciplinary Perspective', *Geographical
    Review*, *59*, 417–26
Buxton, M. and Klein, R. (1978) *Allocating Health Resources: A Commentary on
    the Report of the Resource Allocation Working Party*, Research Paper no. 3,
    Royal Commission on the NHS, HMSO
Byers, T. and Nolan, P. (1976) *Inequality: India and China Compared*, Open
    University, Course D302
Campbell, A.V. (1978) *Medicine, Health and Justice*, Churchill-Livingstone
Carlson, R.J. (1975) *The End of Medicine*, Wiley
Carrier, J. and Kendall I. (1977) 'The Development of Welfare States: The
    Production of Plausible Accounts', *Journal of Social Policy*, *6*, 271–90
Carstairs, V. and Patterson, P.E. (1966) 'Distribution of Hospital Patients by
    Social Class', *Health Bulletin*, *24*, 59–65
Cartwright, A. and O'Brien, M. (1976) 'Social Class Variations in Health Care' in
    M. Stacey (ed), *The Sociology of the N.H.S.*, Sociological Review Monograph,
    22, University of Keele
Cartwright, F.E. (1977) *A Social History of Medicine*, Longman
Castle, I. and Gittus, E. (1957) 'The Distribution of Social Defects in Liverpool'
    Sociological Review, *5*, 43–64
Chadwick, E. (1965) *Report on the Sanitary Condition of the Labouring Population
    of Great Britain*, Edinburgh UP
City and East London AHA(T) (1978) *Area Strategic Plan 1978–88*, City and East

London AHA(T)

Clarke-Kennedy, A.E. (1979) *London Pride. The Story of a Voluntary Hospital*, Hutchinson-Benham

Coates, B.E. and Rawstron, E.M. (1971) *Regional Variations in Britain*, Batsford

Cobb, S. (1976) 'Social Support as a Moderator of Life Stress', *Psychosomatic Medicine*, *38*, 300–314

Cochrane, A.L. (1972) *Effectiveness and Efficiency: Random Reflections on the Health Service*, Nuffield Provincial Hospitals Trust

Cochrane, A.L. St Leger, A. and Moore, F. (1978) 'Health Service Inputs and Mortality Output in Developed Countries', *Journal of Epidemiology and Community Health*, *32*, 200–5

Cockburn, C. (1977) *The Local State*, Pluto Press

Cohen, H. (1961) 'The Evolution of the Concept of Disease' in B. Lush (ed), *Concepts of Medicine*, Pergamon Press

Cole, J.P. and Harrison, M.E. (1978) 'Regional Inequality in the Availability of Services and Purchasing Power in the U.S.S.R., 1940–1976', *Queen Mary College, Department of Geography, Occasional Paper, 14*

Collard, D. (1978) *Altruism and Economy*, Martin Robertson

College of General Practitioners (1961) 'Chronic Bronchitis in Great Britain', *British Medical Journal*, *2*, 973–9

Collins, E. and Klein, R. (1980) 'Equity and the N.H.S.: Self Reported Morbidity, Access and Primary Care', *British Medical Journal*, *281*, 1111–15

Coman, P. (1977) *Catholics and the Welfare State*, Longman

Community Development Project (1977a) *Gilding the Ghetto*, Inter-project Team
———— (1977b) *The Costs of Industrial Change*, Inter-project Team

Conrad, P. and Schneider, J. (1980) *Deviance and Medicalisation: From Badness to Sickness*, C.V. Mosby

Cooke, P. (1980) 'Capital Relation and State Dependency: An Analysis of Urban Development in Cardiff' in F. Rees and T.L. Rees (eds.), *Poverty and Social Inequality in Wales*, Croom Helm

Coombe, V. (1976) 'Health and Social Services and Minority Ethnic Groups', *Royal Society of Health Journal*, *96*, 34–8

Cooper, M.H. (1975) *Rationing Health Care*, Croom Helm

Cordle, F. and Tyroler, H.A. (1974) 'The Use of Medical Records for Epidemiological Research. I. Differences in Hospital Utilization and in Hospital Mortality by Age-Race-Sex-Place of Residence and Socioeconomic Status in a Defined Community Population', *Medical Care*, *XII*, 596–610

Coupland, V. (1982) Gender, Class and Space as Accessibility Constraints for Women with Young Children in 'Contemporary Perspectives on Health Care', Health Research Group, *Queen Mary College, Department of Geography, Occasional Paper, 20*

Cowan, P. (1967) 'Hospital Systems and Systems of Hospitals', *Transactions of the Bartlett Society*, *5*, 101–22

Cox, K. (1965) 'The Application of Linear Programming to Geographical Problems', *TESG*, *56*, 228–36

Crawford, R. (1980) 'Healthism and the Medicalisation of Everyday Life', *International Journal of Health Services*, *10*, 365–88

Croizier, R.C. (1970) 'Medicine, Modernization and Cultural Crisis in China and India', *Comparative Studies in Social History*, *12*, 275–91

Culyer, A.J. (1976) *Need and the N.H.S.*, Martin Robertson

Culyer, A.J. Lavers, R.J. and Williams, A. (1972) 'Health Indicators' in A. Shonfield and S. Shaw (eds), *Social Indicators and Social Policy*, Heinemann

Danielson, R. (1981) 'Medicine in the Community: The Ideology and Substance of Medicine in Socialist Cuba', *Social Science and Medicine*, *15C*, 239–47

Darke, J. and Darke, R. (1970) 'Health and Environment: High Flats', *Centre for Environmental Studies, University Working Paper, 10*

Deitch, R. (1982) 'Commentary from Westminster: Fate of Community Health Councils', *The Lancet, 1* (8243), 116–17

Department of Community Medicine, Middlesex Hospital Medical School, Westminster Medical School, and the Department of Epidemiology, St Mary's Hospital Medical School (1978) *A Study of Attendances at Outpatient Departments in Hospitals in Central London and South Bedfordshire*

Department of Health and Social Security (DHSS) (1980) *Inequalities in Health: Report of a Research Working Group*, Chairman Sir Douglas Black, DHSS

—— (1981a) *Primary Health Care in Inner London: Report of a Study Group*, The London Health Planning Consortium, DHSS

—— (1981b) *'Health Service Development: Community Health Councils'*, Health Circular, *81*, 15, DHSS

—— (1982) *'Health Service Development: Professional Advisory Machinery'*, Health Circular, HC(82), 1

Dever, G.E.A. (1972a) 'Leukemia in Atlanta, Georgia', *Southeastern Geographer, 12*, 91–100

—— (1972b) 'Leukemia and Housing: An Intra-urban Analysis' in N. McGlashan (ed.), *Medical Geography*, Methuen

De Visé, P. (1973) *Misused and Misplaced Hospitals and Doctors: A Locational Analysis of the Urban Health Care Crisis*, Association of American Geographers, Resource Paper No. 22

Dewdney, J.C.H. (1972) *Australian Health Services*, John Wiley and Sons

Dingwall, R. (1976) *Aspects of Illness*, Martin Robertson

Dohrenwend, B.P. and Dohrenwend, B.S. (1969) *Social Status and Psychological Disorder*, Wiley

Donnison, D. (1975) 'Equality', *New Society*, 20 November, 422–4

Dore, R. (1973) *British Factory: Japanese Factory*, Allen and Unwin

Doyal, L. (1979a) 'A Matter of Life and Death: Medicine, Health and Statistics' in J. Irvine, I. Miles and J. Evans (eds), *Demystifying Social Statistics*, Pluto Press

—— (1979b) *The Political Economy of Health*, Pluto Press

Dubos, R. (1960) *Mirage of Health*, Allen and Unwin

—— (1963) 'Infection into Disease' in D.J. Ingle (ed.), *Life and Disease*, Basic Books

Dunbar, J. and Stunkard, A. (1979) 'Adherence to Diet and Drug Regimen' in R. Levy *et al.* (eds), *Nutrition, Lipids and Coronary Heart Disease*, Raven Press

Dunham, H.W. (1937) 'The Ecology of Functional Psychoses in Chicago' *American Sociological Review, 2*, 467–79

Durkheim, E. (1938) *Rules of Sociological Method*, University of Chicago Press

—— (1964) *Suicide*, Routledge and Kegan Paul

Ehrenreich, B. and English, D. (1978) 'The "Sick" Women of the Upper Classes' in J. Ehrenreich (ed.), *The Cultural Crisis of Modern Medicine*, Monthly Review Press

Ehrenreich, J. (ed.) (1978) *The Cultural Crisis of Modern Medicine*, Monthly Review Press

Eisenberg, L. and Kleinman, A. (1981) *The Relevance of Social Science for Medicine*, D. Reidd

Elling, R.H. (1977) 'Industrialisation and Occupational Health in the U.D.Cs', *International Journal of Health Services, 7*, 209–35

—— (1981) 'The Fiscal Crisis of the State and State Financing of Health Care', *Social Science and Medicine, 15C*, 207–17

Elston, M. (1977) 'Medical Autonomy: Challenge and Response' in K. Barnard and K. Lee (eds), *Conflicts in the N.H.S.*, Croom Helm

Engel, G.L. (1963) 'A Unified Concept of Health and Disease' in D.J. Ingle (ed.), *Life and Disease*, Basic Books

Engels, F. (1969) *The Condition of the Working Class in England*, Panther

Escudero, J.C. (1981) 'Democracy, Authoritarianism and Health in Argentina', *International Journal of Health Services*, *11*, 559–72

Evans-Pritchard, E.E. (1937) *Witchcraft, Oracles and Magic Among the Azande*, Clarendon Press

Eyles, J. (1979) 'Area-based Policies for the Inner City' in D.T. Herbert and D.M. Smith (eds), *Social Problems and the City*, Oxford UP

—— (1981a) 'Why Geography Cannot be Marxist: Towards an Understanding of Lived Experience' *Environmental Planning*, *13A*, 1371–88

—— (1981b) 'Ideology, Contradiction and Struggle', *Antipode*, *13*, 39–46

—— (1982) Health and Medicine in Urban Society: The Social Construction and Fetishism of Health in Health Research Group (ed.), 'Contemporary Perspectives on Health and Health Care', *Queen Mary College, Department of Geography, Occasional Paper*, *20*

Eyles, J. and Smith, D.M. (1978) 'Social Geography', *American Behavioural Scientist*, *22*, 41–58

Eyles, J. Smith, D.M. and Woods, K.J. (1982) 'Spatial Resource Allocation and State Practice: The Case of Health Service Planning in London', *Regional Studies*, *16*, 239–53

Fabrega, N. (1975) 'The Need for an Ethnomedical Science', *Science*, *189*, 969–75

—— (1976) 'Towards a Theory of Human Disease', *Journal of Nervous and Mental Disease*, *162*, 299–312

Fanning, D.M. (1967) 'Families in Flats', *British Medical Journal*, *4*, 382–6

Faris, R.E.L. and Dunham, H.W. (1939) *Mental Disorders in Urban Areas*, University of Chicago Press

Fawcett, J.T. (1970) *Psychology and Population*, The Population Council

Fay, B. (1975) *Social Theory and Political Practice*, Allen and Unwin

Fenton Lewis, A. and Modle, W.J. (1982) 'Health Indicators: What are They? An Approach to Efficacy in Health Care', *Health Trends*, *14*, 3–8

Field, M.G. (1967) *Soviet Socialised Medicine*, Collier MacMillan

—— (1976) 'Health as a Public Utility or the Maintenance of Capacity in Soviet Society' in M.G. Field (ed.), *Social Consequences of Modernization in Communist Countries*, Johns Hopkins UP

Forest, D. and Sims, P. (1982) 'Health Advisory Services and the Immigrant', *Health Trends*, *14*(1), 10–13

Forster, D.P. (1977) 'Mortality, Morbidity and Resource Allocation', *Lancet*, *8019*, 997–8

Forsyth, G. (1973) *Doctors and State Medicine*, Pitman Medical

Fosu, G.B. (1981) 'Disease Classification in Rural Ghana', *Social Science and Medicine*, *15B*, 471–82

Foucault, M. (1973) *The Birth of the Clinic*, Tavistock

Fox, P.D. (1978) 'Managing Health Resources: English Style' in G. McLachlan (ed.), *By Guess or By What? Information Without Design in the N.H.S.*, Oxford UP

Frankenberg, R. (1974) 'Functionalism and After?', *International Journal of Health Services*, *4*, 411–27

—— (1981) 'Allopathic Medicine, Profession and Capitalist Ideology in India', *Social Science and Medicine*, *15B*, 115–25

Freidson, E. (1960) 'Client Control and Medical Practice', *American Journal of Sociology*, *65*, 374–82

—— (1970a) *Profession of Medicine*, Dodd Head

—— (1970b) *Professional Dominance*, Aldine

Fried, A. and Elman, R. (1969) *Charles Booth's London*, Hutchinson
Fried, M. (1963) 'Grieving for a Lost Home' in L.J. Duhl (ed.), *The Urban Condition*, Basic Books
Gans, H.J. (1962) *The Urban Villagers*, Free Press
Gardner, M.J. Winter. P.D. and Acheson, E.D. (1982) 'Variation in Cancer Mortality Areas in England and Wales: Relations with Environmental Factors and Search for Causes', *British Medical Journal*, *284*, 784–7
Geary, K. (1977) 'Technical Deficiencies in RAWP', *British Medical Journal*, 21 May, 1
Gibbons, J. (1978) 'The Mentally Ill' in P. Brearley, J. Gibbons, A. Miles, E. Topliss and G. Woods, *The Social Context of Health Care* Basil Blackwell and Martin Robertson
Gibson, J.B. and Johansen, A. (1979) *The Quick and the Dead: A Biomedical Atlas of Sydney*, Reed
Giddens, A. (1973) *The Class Structure of the Advanced Societies*, Hutchinson
Giggs, J.A. (1970) 'Socially Disorganised Areas in Barry' in H. Carter and W.K.D. Davies (eds), *Urban Essays: Studies in the Geography of Wales*, Longman
——— (1973) 'The Distribution of Schizophrenics in Nottingham', *Transactions of the Institute of British Geographers*, *59*, 55–76
——— (1979) 'Human Health Problems in Urban Areas' in D.T. Herbert and D.M. Smith (eds), *Social Problems and the City*, Oxford UP
Giggs, J.A. Ebdon, D.S. and Bourke, J.B. (1980) 'The Epidemiology of Primary Acute Pancreatitis in the Nottingham Defined Population Area', *Trans. IBG*, *5*, 229–42
Girt, J.L. (1972) 'Simple Chronic Bronchitis and Urban Ecological Structure' in N. McGlashan (ed.), *Medical Geography*, Methuen
——— (1974) 'The Geography of Non-vectored Infectious Diseases' in J.M. Hunter (ed.), *The Geography of Health and Disease*, University of North Carolina, Department of Geography
Glass, R. (1968) 'Urban Sociology in Great Britain' in R. Pahl (ed.), *Readings in Urban Sociology*, Pergamon
Glick, L.B. (1967) 'Medicine as an Ethnographic Category: The Gimi of the New Guinea Highlands', *Ethnology*, *6*, 31–56
Godlund, S. (1961) 'Population, Regional Hospitals, Transport Facilities and Regions: Planning the Location of Regional Hospitals in Sweden', *Lund Studies in Geography, Series B, Human Geography*, *21*, 3–32
Goldberg, E.M. and Morrison, S.L. (1963) 'Schizophrenia and Social Class', *British Journal of Psychiatry*, *109*, 785–802
Goldstein, M.S. and Donaldson, P.J. (1979) 'Exploiting Professionalism: A Case Study of Medical Education', *Journal of Health and Social Behaviour*, *20*, 322–37
Gough, I. (1979) *The Political Economy of the Welfare State*, Macmillan
Gould, P.R. and Leinbach, T.R. (1966) 'An Approach to the Geographic Assignment of Hospital Services', *TESG*, *57*, 203–6
Gravelle, H.S.E. Hutchinson, G. and Stern, J. (1981) 'Mortality and Unemployment: A Critique of Brenner's Time-Series Analysis', *The Lancet*, 26 September, 675–81
Gt Britain (1962) *A Hospital Plan for England and Wales*, Ministry of Health, Cmnd 1604, HMSO
——— (1969) *The Functions of the District General Hospital*, Report of the Committee (Chairman, Sir Desmond Bonham-Carter), Central Health Services Council, HMSO
——— (1971) *Better Services for the Mentally Handicapped*, Cmnd 4683, HMSO
——— (1975) *Better Services for the Mentally Ill*, Cmnd 6233, HMSO

——— (1976a) *Priorities for Health and Personal Social Services in England*, HMSO

——— (1976b) *Sharing Resources for Health in England*, Report of the Resource Allocation Working Party, HMSO

——— (1977) *The Way Forward*, HMSO

——— (1978) *The Management of Finance in the NHS*, Research Paper no. 2, Royal Commission on the NHS, HMSO

——— (1979a) *Acute Hospital Services in London: A Profile by the London Health Planning Consortium*, HMSO

——— (1979b) *Royal Commission on the N.H.S: Report*, Chairman Sir Alec Merrison, Cmnd 7615, HMSO

——— (1980) Health and Personal Social Services Statistics, HMSO

——— (1981) *Care in Action*, A Handbook of Policies and Priorities for the Health and Personal Social Services in England, HMSO

Gronholm, L. (1960) 'The Ecology of Social Disorganisation in Helsinki', *Acta Sociologia, 5*, 31–41

Gudgin, G. (1975) 'The Distribution of Schizophrenics in Nottingham: A Comment', *Transactions of the Institute of British Geographers, 64*, 148–9

Guevara, C. (1968) *Socialism and Man in Cuba*, Stage One

Güse, H.G. and Schmake, N. (1980) 'Psychiatry and the Origins of Nazism', *International Journal of Health Services, 10*, 177–96

Guttmacher, S. and Danielson, R. (1977) 'Changes in Cuban Health Care', *International Journal of Health Services, 7*, 383–400

Habermas, J. (1976) *Legitimation Crisis*, Heinemann

Haggett, P. (1976) 'Hybridising Alternative Models of an Epidemic Diffusion Process', *Economic Geography, 52*, 136–46

Hall, P. (1957) *The Social Services of Modern England*, RKP

Hamnett, M.P. and Connell, J. (1981) 'Diagnosis and Cure: The Resort to Traditional and Modern Medical Practitioners in the North Solomons', *Social Science and Medicine, 15B*, 489–98

Hare, E.H. (1956) 'Mental Illness and Social Conditions in Bristol', *British Journal of Preventative and Social Medicine, 9*, 191–5

Hare, E.H. and Shaw, G.K. (1965) *Mental Health on a New Housing Estate*, Oxford UP

Harrington, M. (1962) *The Other America*, Macmillan

Hart, J.T. (1971) 'The Inverse Care Law', *Lancet*, i, 405–12

Hartung, F.E. (1963) 'Manhattan Madness: The Social Movement of Mental Illness', *Sociological Quarterly, 4*, 261–72

Hartwell, R.M. (1971) *The Industrial Revolution and Economic Growth*, Methuen

Harvey, D. (1973) *Social Justice and the City*, Arnold

Hay, J.R. (1975) *The Origins of the Liberal Welfare Reforms, 1906–14*, Macmillan

Haynes, R.M. and Bentham, C.G. (1979) *Community Hospitals and Rural Accessibility*, Saxon House

Haywood, S. and Alaszewski, A. (1980) *Crisis in the Health Service*, Croom Helm

Heclo, H. (1972) 'Policy Analysis', *British Journal of Political Science, 2*, 83–108

Heller, T. (1978) *Restructuring the Health Service*, Croom Helm

Herbert, D.T. (1972) *Urban Geography*, David and Charles

Herbert, D.T. and Peace, S.M. (1980) 'The Elderly in an Urban Environment' in D.T. Herbert and R.J. Johnston (eds), *Geography and the Urban Environment*, Wiley

Hetzel, B.S. (1980) *Health and Australian Society*, Penguin

Higgins, J. (1981) *States of Welfare*, Basil Blackwell and Martin Robertson

Hinkle, L.E. and Wolfe, H.G. (1957) 'Health and the Social Environment' in A.H. Leighton, J.A. Clausen and R.N. Wilson (eds), *Explorations in Social*

Psychiatry, Basic Books
Hirsch, F. (1977) *The Social Limits to Growth*, RKP
Hirsch, J. (1978) 'The State Apparatus and Social Reproduction: Elements of a
  Theory of the Bourgeois State' in J. Holloway and S. Picciotto *State and
  Capital*, Arnold
Hirt, J.F.B. (1966) 'Planning for a New Community', *Journal of the College of
  General Practitioners, 12*, Suppl. no. 1, 33–4
Hobsbawn, E. (1969) *Industry and Empire*, Penguin
Howe, G.M. (1960) 'The Geographical Distribution of Cancer Mortality in Wales
  1947–53', *Trans. IBG, 28*, 190–210
—— (1972) *Man, Environment and Disease in Britain*, David and Charles
—— (1976) 'The Geography of Disease' in C.O. Carter and J. Peel (eds),
  Equalities and Inequalities in Health, Academic Press
Hughes, C. and Hunter, J. (1971) 'Disease and Development in Africa' in H.P.
  Dreitzel (ed.), *The Social Organisation of Health*, Macmillan
Hyde, G. (1974) *The Soviet Health Service*, Lawrence and Wishart
Illich, I. (1974) 'Medical Nemesis', *Lancet, 1*(7863), 918–21
—— (1975) *Medical Nemesis*, Calder and Boyars
—— (1977) *Limits to Medicine*, Penguin
Illsley, R. (1980) *Professional or Public Health?*, The Nuffield Provincial Hospitals
  Trust
Ingram, D.R. Clarke, D.R. and Murdie, R.A. (1978) 'Distance and the Decision
  to Visit an Emergency Department', *Social Science and Medicine, 12*, 55–62
Janzen, J. (1978) 'Micro and Macro-analysis in the Comparative Study of Medical
  Systems', *Social Science and Medicine, 12B*, 121–9
Jephcott, P. (1971) *Homes in High Flats*, Oliver and Boyd
Johnson, A.L. *et al.*, (1962) 'Epidemiology of Polio Vaccine Acceptance', *Florida
  State Board of Health, Monograph, 3*
Jolly, R. and King, M. (1966) 'The Organization of Health Services' in M. King
  (ed.), *Medical Care in Developing Countries*, Oxford UP, Nairobi
Jones, E. and Eyles, J. (1977) *An Introduction to Social Geography*, Oxford UP
Jordan, B. (1974) *Poor Parents*, RKP
Kaim-Caudle, P. (1973) *Comparatively Social Policy and Social Security*, Martin
  Robertson
Kaplan, B.H. Cassel, J. and Gore, S. (1977) 'Social Support and Health', *Medical
  Care, 15*, Suppl. 5, 47–58
Kaser, M. (1976) *Health Care in the Soviet Union and Eastern Europe*, Croom
  Helm
Kelman, S. (1975) 'The Social Nature of the Definition Problem in Health',
  International Journal of Health Services, 5, 625–42
Kennedy, I. (1980) 'Unmasking medicine', *The Listener*, 6 November, 600–4
Kincaid, J.C. (1973) *Poverty and Equality in Britain*, Penguin
Klein, R. and Lewis, J. (1976) *The Politics of Consumer Representation. A Study of
  Community Health Councils*, Centre for Studies in Social Policy
Knight, C.G. (1974) 'The Geography of Vectored Diseases' in J.M. Hunter (ed.),
  The Geography of Health and Disease, University of North Carolina,
  Department of Geography
Knox, P.L. (1978) 'The Intra-urban Ecology of Primary Medical Care: Patterns of
  Accessibility and Their Policy Implications', *Environment and Planning*, A, *10*,
  415–35
—— (1979) 'Medical Deprivation, Area Deprivation and Public Policy', *Social
  Science and Medicine, 13D*, 111–21
Kohn, R. and White, K.L. (1976) *Health Care: An International Study*, Oxford UP
Koos, E.L. (1964) *The Health of Regionville*, Columbia UP

Korsch, B. and Negrete, V. (1972) 'Doctor-patient Communication' *Scientific American*, 227, 66–74

Krause, E. (1977) *Power and Illness*, Elsevier

Kuhn, H.W. and Kuenne, R.E. (1962) 'An Efficient Algorithm for the Solution of the Generalized Weber Problem', *Journal of Regional Science*, 4, 21–33

Kunitz, S. and Levy, J.E. (1974) 'Changing Ideas of Alcohol Use Among Navaho Indians', *Quarterly Journal of Studies in Alcoholism*, 35.

Laing, R.D. (1960) *The Divided Self*, Tavistock

Lall, S. (1977) 'Medicine and Multinationals', *Monthly Review*, 28

The Lancet (1975) 'Round the World: Australia' *The Lancet*, 14 June, 1332

Lasker, J.N. (1981) 'Choosing Among Therapies: Illness Behaviour in the Ivory Coast', *Social Science and Medicine*, 15A, 157–68

Laurell, A.C. (1981) 'Mortality and Working Conditions in Agriculture in the U.D.Cs', *International Journal of Health Services*, 11, 3–20

Learmonth, A.T.A. (1957) 'Some Contrasts in the Regional Geography of Malaria in India and Pakistan', *Trans. IBG*, 23, 32–59

——— (1975) 'Ecological Medical Geography', *Progress in Geography*, 7, 202–26

Lee, R.P.L. (1981) 'Chinese and Western Medical Care in China's Rural Commune', *Social Science and Medicine*, 15B, 137–48

Lees, D.S. (1961) *Health Through Choice: An Economic Study of the British N.H.S.*, Institute of Economic Affairs

Leiss, W. (1978) *The Limits to Satisfaction*, Marion Boyers

Leslie, C. (1974) 'The Modernization of Asian Medical Systems' in J. Poggie and R. Lynch (eds), *Rethinking Modernization*, Greenwood Press

Levitt, R. (1976) *The Reorganized N.H.S.*, Croom Helm

Levy, L. and Rowitz, L. (1973) *The Ecology of Mental Disorders*, Behavioural Publications

Lisitsin, Y. (1972) *Health Protection in the U.S.S.R.*, Progress Publishers

Littlewood, R. and Lipsedge, M. (1982) *Aliens and Alienists*, Penguin

Logan, W.P.D. and Cushion, A.A. (1960) 'Morbidity Statistics from General Practice' in *Studies on Medical and Population Statistics*, HMSO

London Health Planning Consortium (LHPC) (1980) *Towards a Balance: A Framework for Acute Hospital Services in London Reconciling Service With Teaching Needs*. A discussion document issued by the LHPC

Lukes, S. (1973) *Emile Durkheim*, Allen Lane

McGlashan, N. (1966) 'Geographical Evidence on Medical Hypothesis', *Tropical and Geographical Medicine*, 19, 333–43

——— (1969) 'The African Lymphoma in Central Africa', *International Journal of Cancer*, 4, 113–20

——— (1972) *Medical Geography: Techniques and Field Studies*, Methuen

——— (1982) 'Environments for Disease: Oesophageal Cancer', *Geographical Magazine*, 54, 263–7

McHarg, I.L. (1969) *Design with Nature*, Natural History Press

MacKenzie, W.J.M. (1979) *Power and Responsibility in Health Care*, Oxford UP for the Nuffield Provincial Hospitals Trust

McKeown, T. (1971) 'A Historical Appraisal of the Medical Task' in G. McLachlan and T. McKeown (eds), *Medical History and Medical Care*, Oxford UP

——— (1976) *The Role of Medicine – Dream, Mirage or Nemesis?*, Nuffield Provincial Hospitals Trust

——— (1979) *The Role of Medicine*, Nuffield Provincial Hospitals Trust

McKeown, T. and Lowe, C.R. (1966) *An Introduction to Social Medicine*, F.A. Davies

Madison, B.Q. (1968) *Social Welfare in the Soviet Union*, Stanford UP

Mandel, E. (1975) *Late Capitalism*, Verso
Manzie, P.P. (1979) 'The Rapture of China', *Medical Journal of Australia*, 30 June, 601–2
Maris, R.W. (1969) *Social Forces in Urban Suicide*, Dorsey Press
Marshall, T.H. (1965) *Class, Citizenship and Social Development*, Doubleday Anchor
—— (1975) *Social Policy*, Hutchinson
Martin, F.M. Brotherstone, J.H.F. and Chave, S.P.W. (1957) 'Incidence of Neurosis in a New Housing Estate', *British Journal of Preventative Social Medicine*, *11*, 196–202
Marx, K. *Capital*, Vol. 1, Lawrence and Wishart
—— *Grundrisse*, Penguin
Marx, K. and Engls, F. (1970) *The German Ideology*, Lawrence and Wishart
May, J.M. (1950) 'Medical Geography: Its Methods and Objectives', *Geographical Review*, *40*, 9–41
—— (1961) *Studies in Disease Ecology*, Hafner Publishing Co.
Mechanic, D. (1978) *Medical Sociology*, Free Press
Mercer, C. (1975) *Living in Cities: Psychology and the Urban Environment*, Penguin
Mills, C. (1944) *Climate Makes the Man*, Gollancz
Mintz, N.L. and Schwarz, D.T. (1964) 'Urban Ecology and Psychosis', *International Journal of Social Psychiatry*, *10*, 101–18
Mishler, E.G. (1981) 'Critical Perspectives on the Biomedical Model' in E.G. Mishler *et al.* (eds), *Social Contexts of Health, Illness and Patient Care*, Cambridge UP
Mishra, R. (1977) *Society and Social Policy*, Macmillan
Moore, N.C. (1974) 'Psychiatric Illness and Living in Flats', *British Journal of Psychiatry*, *125*, 500–7
Morrill, R.L. and Earickson, R. (1969a) 'Locational Efficiency of Chicago Hospitals: An Experimental Model', *Health Services Research*, *4*, pt 2, 128–41
Morrill, R.L. and Earickson, R. (1969b) 'Problems of Modelling Interaction: the Case of Hospital Care', in K.L. Cox and R.G. Golledge (eds.) *Behavioural Problems in Geography*, Northwestern University, Department of Geography, Research Studies 17
Morrill, R.L. and Kelley, M.B. (1970) 'The Simulation of Hospital Use and the Estimation of Locational Efficiency', *Geographical Analysis*, *2*, 283–300
Morris, J.N. (1975) *Uses of Epidemiology*, Livingston
Morris, J.N. and Heady, J.A. (1955) 'Social and Biological Factors in Infant Mortality: V. Mortality in Relation to Father's Occupation 1911–50', *The Lancet*, i, 554–9
Morris, R.N. (1965) *Urban Sociology*, Allen and Unwin
Navarro, V. (1976) *Medicine Under Capitalism*, Prodist
—— (1978) *Class Struggle, the State and Medicine*, Martin Robertson
North East Thames Regional Health Authority (NETRHA) (1978) Strategic Plan 1978–1988 (revised draft), NETRHA, Eastbourne Terrace, London W2
O'Connor, J. (1973) *The Fiscal Crisis of the State*, St Martin's Press
Offe, C. (1972) 'Political Authority and Class Structures: An Analysis of Late Capitalist Societies', *International Journal of Sociology*, *2*, 73–105
Oppé, T.E. (1964) 'The Health of West Indian Children', *Proc. Roy. Soc. Med.*, *57*, 321–3
O'Sullivan, D. (1981) ''Health Care Plans in Australia', *Benefits International*, *11*(4), 7–9
Owen, D. (1976) *In Sickness and in Health*, Quarter Books
Page, E. (1960) *What Price Medical Care: A Preventative Prescription for Private*

Medicine, J.B. Lippincott
Park, R.E. (1967) 'The City: Suggestions for the Investigation of Human Behaviour in the Urban Environment' in R.E. Park and E.W. Burgess, *The City*, University of Chicago Press
Parker, J. (1975) *Social Policy and Citizenship*, Macmillan
Parkin, F. (1971) *Class Inequality and Political Order*, McGibbon and Kee
Parsons, T. (1951) *The Social System*, Free Press
——— (1964) *Social Structure and Personality*, Free Press
Pfeil, E. (1968) 'The Pattern of Neighbouring Relations in Dortmund-Nordstadt' in R.E. Pahl (ed.), *Readings in Urban Sociology*, Pergamon Press
Phillips, D.R. (1979) 'Spatial Variations in Attendance at General Practitioner Services', *Social Science and Medicine*, *13D*, 169–81
——— (1981) *Contemporary Issues in the Geography of Health Care*, Geobooks
Pilisuk, M. and Froland, C. (1978) 'Kinship, Social Networks, Social Support and Health', *Social Science and Medicine*, *12B*, 273–80
Pinker, R. (1971) *Social Theory and Social Policy*, Heinemann
Piven, F.F. and Cloward, R.A. (1970) *Regulating the Poor: The Functions of Social Welfare*, Tavistock
Poggi, G. (1978) *The Development of the Modern State*, Hutchinson
Potts, D.M. (1976) 'Problems and Solutions in Developing Countries' in C.O. Carter and J. Peel (eds), *Equalities and Inequalities in Health*, Academic Press
Poulantzas, N. (1973) *Political Power and Social Classes*, NLB
Powles, J. (1973) 'On the Limitations of Modern Medicine', *Science, Medicine and Man*, *1*, 1–30
Pyle, G.F. (1973) 'Measles as an Urban Health Problem: The Akron Example' *Economic Geography*, *49*, 344–356
——— (1976) 'Foundations to Medical Geography', *Economic Geography*, *52*, 95–102
——— (1977) 'International Communication and Medical Geography', *Social Science and Medicine*, *11*, 679–82
——— (1979) *Applied Medical Geography*, Winston-Wiley
Pyle, G.F. and Lauer, B.M. (1975) 'Comparing Spatial Configurations: Hospital Service Areas and Disease Rates', *Economic Geography*, *51*, 50–68
Pyle, G.F. and Rees, P.H. (1971) 'Modelling Patterns of Death and Disease in Chicago', *Economic Geography*, *47*, 475–88
Raphael, D.D. (1970) *Problems of Political Philosophy*, Macmillan
Redlich, F.C. (1957) 'The Concept of Health in Psychiatry' in A.H. Leighton, J.N. Clausen and R.N. Wilson (eds), *Explorations in Social Psychiatry*, Tavistock
Rein, M. (1969) 'Social Class and the Utilization of Medical Care Services', *Journal of the American Hospital Association*, *43*, 43–54
Revelle, C. Marks, D. and Leibmann, J.C. (1970) 'An Analysis of Private and Public Sector Location Models', *Management Science*, *16*, 692–707
Rhodes, P. (1976) *The Value of Medicine*, Allen and Unwin
Richman, N. (1974) 'The Effects of Housing on Pre-school Children and Their Mothers', *Develop. Med. Chil. Neurol.*, *16*, 53–8
Rickard, J.E. (1976) '*Per Capita* Expenditure of English Area Health Authorities', *British Medical Journal*, 31 January, 299–300
Ritter, A.R.M. (1974) *The Economic Development of Revolutionary Cuba*, Praeger
Roberts, R. (1971) *The Classic Slum*, University of Manchester Press
Roberts, D.S. (1976) 'Sex Differences in Disease and Mortality' in C.O. Carter and J. Peel (eds), *Equalities and Inequalities in Health*, Academic Press, 13–34
Robinson, D. (1971) *The Process of Becoming Ill*, Routledge and Kegan Paul
Rodgers, B. (1968) *The Battle Against Poverty: Vol. 1 – From Pauperism to Human*

Rights, RKP

Room, G. (1979) *The Sociology of Welfare,* Martin Robertson

Roszak, T. (1972) *Where the Wasteland Ends,* Doubleday

Roundy, R.W. (1976) 'Altitudinal Mobility and Disease Hazards for Ethiopian Populations', *Economic Geography, 52,* 103–15

Rowntree, B.S. (1901) *Poverty: A Study of Town Life,* Macmillan

Rowntree, B.S. and Lavers, G.R. (1951) *Poverty and the Welfare State,* Longman

Rushton, G. Goodchild, M. and Ostresh, S. (1973) 'Computer Programs for Location-Allocation Problems', *University of Iowa, Department of Geography, Monograph 6*

Rutter, M. and Madge, N. (1976) *Cycles of Disadvantage,* Heinemann

Ryan, M. (1978) *The Organization of Medical Care in the Soviet Union,* Basil Blackwell and Martin Robertson

Ryle, J. (1961) 'The Meaning of Normal' in B. Lush (ed.), *Concepts of Medicine,* Pergamon Press

Sainsbury, P. (1955) *Suicide in London,* Clapman and Hall

St Thomas' Health District (1978) *Survey of Outpatient Flows into St Thomas' Health District,* St Thomas' Health District, London

Saul, S.B. (1969) *The Myth of the Great Depression 1873–1896,* Macmillan

Sayer, D. (1979) *Marx's Method: Ideology, Science and Critique in 'Capital',* Harvester Press

Scarpa, A. (1981) 'Pre-scientific Medicines: Their Extent and Value', *Social Science and Medicine, 15A,* 317–26

Schmitt, R.C. (1963) 'Implications of Density in Hong Kong', *Journal of the American Institute of Planners, 29,* 210–17

—— (1966) 'Density, Health and Social Organisation', *Journal of the American Institute of Planners, 32,* 38–40

Schneider, J.B. and Symons, J.G. (1971) *Regional Health Facility System Planning: An Access Opportunity Approach,* Research Science Research Institute, Discussion Papers *48*

Schur, E.M. (1971) *Labelling Deviant Behaviour,* Harper and Row

Scotton, R.B. (1974) *Medical Care in Australia: An Economic Diagnosis,* Sun Books, for the Institute of Applied Economic and Social Research, University of Melbourne

Shannon, G.W. (1980) 'The Utility of Medical Geography Research', *Social Science and Medicine, 14D,* 1–2

Shannon, G.W. and Dever, G.E.A. (1974) *The Geography of Health Care,* McGraw-Hill

—— (1979) *Health Care Delivery: Spatial Perspectives,* McGraw-Hill

Shaw, G.B. (1977) *The Doctor's Dilemma,* Penguin

Shoskin, A.A. (1964) 'Geographical Aspects of Public Health', *Soviet Geography, 5,* 72–8

Shuval, J.T. Antonovsky, A. and Davies, A.M. (1970) *Social Functions of Medical Practice,* Jossey-Bass

Sidel, R. (1977) 'People Serving People' in D. Thurz and J.L. Vigilante (eds), *Meeting Human Needs,* Sage Publications

Sidel, V.W. and Sidel, R. (1973) *Serve the People – Observations on Medicine in the People's Republic of China,* Beacon Press

Simon, H.A. (1957) *Models of Man,* Wiley

Sinfield, A. (1978) 'Analyses in the Social Division of Welfare', *Journal of Social Policy, 7,* 129–56

Smith, B.E. (1981) 'Black Lung: The Social Production of Disease', *International Journal of Health Services, 11,* 343–59

Smith, D.M. (1973) *The Geography of Social Well-being in the U.S.,* McGraw-Hill

—— (1977) *Human Geography: A Welfare Approach*, Edward Arnold
—— (1979) *Where the Grass is Greener: Living in an Unequal World*, Penguin
—— (1982) 'Geographical Perspectives on Health and Health Care', Health Research Group, *Queen Mary College, Department of Geography, Occasional Paper 20*
Smith, F.B. (1979) *The People's Health 1830–1910*, Croom Helm
Smith, H. (1976) *The Russians*, New York Times Book Company
Smith, M.P. (1980) *The City and Social Theory*, Basil Blackwell
Snaith, A.H. (1978) 'Sub-regional Allocations in the NHS', *Journal of Epidemiology and Community Health, 32*, 16–22
Soboleva, L.I. (1974) 'Medical Geography Aspects of the Design of the Ust-Ilimsk Industrial Node', *Soviet Geography, 15*, 422–8
Sorokina, M.V. (1976) 'The Impact of Environmental Factors on the Incidence of Lung Cancer Among Rural Residents', *Soviet Geography, 17*, 125–9
Sorre, M. (1957) *Rencontres de la Géographie et de la Sociologie*, Editions
Srole, L. (1962) *Mental Health in the Metropolis*, Harper and Row
Stamp, L.D. (1964) *Some Aspects of Medical Geography*, Oxford UP
Stedman Jones, G. (1971) *Outcast London*, Oxford UP
Stewart, J.Q. (1948) 'Demographic Gravitation: Evidence and Applications', *Sociometry, 11*, 31–56
Stimson, G. and Stimson, C. (1978) *Health Rights Handbook*, Prism Press
Stimson, R.J. (1977) 'The Distribution and Utilization of Health Care Services in Metropolitan Adelaide' in C.A. Forster and R.J. Stimson (eds), 'Urban South Australia: Selected Readings', *Flinders University, Centre for Applied Social Sciences and Survey Research, Monograph Series, 1*
—— (1980) 'Spatial Aspects of Epidemiological Phenomena and of the Provision and Utilization of Health Care Services in Australia: A Review of Methodological Problems and Empirical Analyses', *Environment and Planning, A, 12*, 881–907
Suchmann, E.A. (1965) 'Social Patterns of Illness and Medical Care', *Journal of Health and Human Behaviour, 6*, 2–16
Sugimoto, T. (1968) 'The Contribution of Buddhism to Social Work' in D. Dessan (ed.), *Glimpses of Social Work in Japan*, Social Work International
Susser, M.W. and Watson, W. (1971) *Sociology in Medicine*, Oxford UP
Symons, J.G. (1971) 'Some Comments on Equity and Efficiency in Public Facility Location Models', *Antipode, 3(1)*, 54–67
Szasz, T. (1970) *The Manufacture of Madness*, Harper and Row
Tabor, D.C. (1981) 'Ripe and Unripe: Concepts of Health and Sickness in Ayurvedic Medicine', *Social Sciences and Medicine, 15B*, 439–55
Tarn, J.N. (1966) 'The Peabody Donation Fund: The Role of a Housing Society in the Nineteenth Century', *Victorian Studies, 10*, 7–38
Tawney, R.H. (1966) *Religion and the Rise of Capitalism*, Penguin
Taylor, J.S. (1976) 'Problems of Minimum Cost Location: The Kuhn and Kuenne Algorithm', *Queen Mary College, Department of Geography, Occasional Paper, 4*
Taylor, S. and Chave, S. (1964) *Mental Health and Environment*, Little, Brown and Co.
Teitz, N.B. (1968) 'Toward a Theory of Urban Public Facility Location', *Papers of the Regional Science Association, 21*, 35–51
Thomas, H.E. (1968) 'Tuberculosis in Immigrants', *Proc. Roy. Soc. Med., 61*, 21–3
Timms, D.W.G. (1965) 'The Spatial Distribution of Social Deviants in Luton', *Australian and NZ Journal of Sociology, 1*, 38–52
Titmuss, R.M. (1968) *Commitment to Welfare*, Allen and Unwin
—— (1973) *The Gift Relationship*, Penguin

—— (1974) *Social Policy*, Allen and Unwin
—— (1976) *Essays on the Welfare State*, Allen and Unwin
Topping, P. and Smith, G. (1977) *Government Against Poverty? Liverpool Community Development Project 1970–5*, Social Evaluation Unit
Totman, R. (1979) *Social Causes of Illness*, Pantheon Books
Townsend, P. (1954) 'Measuring Poverty', *British Journal of Sociology*, 5, 130–7
—— (1976) 'The Difficulties of Policies Based on the Concept of Area Deprivation', Queen Mary College, Department of Economics, Barnett Shine Foundation Lecture
—— (1979) *Poverty in the United Kingdom*, Penguin
Tyne, A. (1978) *Review of Progress in Provision for Mentally Handicapped People*, CMH
University of London (1980) *Report of a Working Party on Medical and Dental Teaching Resources*, Chairman, The Lord Flowers, University of London
Urry, J. (1981) *The Anatomy of Capitalist Societies*, Macmillan
Vallin, J. (1975) 'La Mortalité en Algerie', *Population*, 30, 1023–46
Voronov, A.G. (1977) 'The Geographical Environment and Human Health', *Soviet Geography*, 18, 230–7
Waitzkin, H. (1981) 'The Social Origins of Illness', *International Journal of Health Services*, 11, 77–103
Waitzkin, H. and Stoeckle, J.D. (1976) 'Information Control and the Micropolitics of Health Care', *Social Science and Medicine*, 10, 263–76
Wallerstein, I. (1974) *The Modern World System*, Academic Press
Walmsley, D.J. (1978) 'The Influence of Distance on Hospital Usage in Rural New South Wales', *Australian Journal of Social Issues*, 13, 72–81
Walters, V. (1980) *Class Inequality and Health Care*, Croom Helm
Wardle, C.J. (1962) 'Social Factors in the Major Functional Psychoses' in A.T. Welford *et al.* (eds), *Society: Problems and Methods of Study*, Routledge and Kegan Paul
Waxler, N.E. (1977) 'Is Mental Illness Cured in Traditional Societies?', *Culture, Medicine and Psychiatry*, 1, 233–53
—— (1981) 'Learning to be a Leper: A Case Study in the Construction of Illness' in E.G. Mishler *et al.* (eds), *Social Contexts of Health, Illness and Patient Care*, Cambridge UP
Webb, P. (1982) 'The Clash of Cultures – Health Care', *Nursing*, 2, 20–2
Wedderburn, D. (1965) 'Facts and Theories About the Welfare State' in R. Miliband and J. Saville (eds), *The Socialist Register*, Merlin Press
Wilenski, P. (1977) *The Delivery of Health Services in the People's Republic of China*, ANU Press
Wilensky, H. and Lebeaux, C. (1965) *Industrial Society and Social Welfare*, Free Press
Wilensky, H. (1976) *The New Corporatism, Centralisation and The Welfare State*, Sage
Williams, A. (1974) '"Need" as a Demand Concept?' in A.J. Culyer (ed.), *Economic Policies and Social Goals*, Martin Robertson
Williams, R. (1973) *The Country and the City*, Chatto and Windus
—— (1977) *Marxism and Literature*, Oxford UP
Wirth, L. (1938) 'Urbanism as a Way of Life', *American Journal of Sociology*, 44, 1–24
Wolff, S. (1973) *Children Under Stress*, Penguin
Woodrofe, K. (1962) *From Charity to Social Work*, RKP
Woods, K.J. (1979) *Social Deprivation and Hospital Utilization in the Tower Hamlets Health District*, Report to the City and East London AHA(T), Department of Geography, Queen Mary College, University of London

—— (1982) Social Deprivation and Resource Allocation in the Thames Regional Health Authorities in Health Research Group (ed.), 'Contemporary Perspectives on Health and Health Care', *Queen Mary College, Department of Geography, Occasional Paper, no. 20*

Woolf, K. (ed.) (1950) *The Sociology of Georg Simmel*, Free Press

World Health Organization (1980) *The Sixth World Health Situation Report 1973–1977: Part I – Global Analysis*, WHO

Wright, E.O. (1978) *Class, Crisis and The State*, New Left Books

Young, G.M. (1936) *Victorian England*, Oxford UP

Zborowski, M. (1952) 'Cultural Components in Responses to Pain', *Journal of Social Issues, 8*, 16–30

Zipf, G.K. (1949) *Human Behaviour and the Principle of Least Effort*, Cambridge, Massachusetts

Zola, I.K. (1966) 'Culture and Symptoms: An Analysis of Patients Presenting Complaints', *American Sociological Review, 21*, 615–30

—— (1972) 'Medicine as an Institution of Social Control', *Sociological Review, 20*, 487–503

Zwick, P. (1976) 'Intrasystem Inequality and the Symmetry of Socioeconomic Development in the U.S.S.R.', *Comparative Politics, 8*, 501–24

SUBJECT INDEX

AUTHOR INDEX

Milton Keynes UK
Ingram Content Group UK Ltd.
UKHW031146141024
449569UK00024B/1037

9 781138 998100